Caribbean Comm

A Social Studies Series for

Saint Lucia

Colin Brock

MACMILLAN
CARIBBEAN

First published 1976
Reprinted 1980, 1982, 1983, 1984, 1985, 1986 (twice), 1987

Published by *Macmillan Publishers Ltd*
London and Basingstoke
*Associated companies and representatives in Accra,
Auckland, Delhi, Dublin, Gaborone, Hamburg, Harare,
Hong Kong, Kuala Lumpur, Lagos, Manzini, Melbourne,
Mexico City, Nairobi, New York, Singapore, Tokyo.*

ISBN 0 333 19780 1

Printed in Hong Kong

Contents

Acknowledgements

The author wishes to express his gratitude to the following people of Saint Lucia for their help in the preparation of this book: to Thomas Daniel of Desrussieux for his advice on aspects of the festivals and carnival; to Mrs Augustin and her family of Garrand for so willingly participating in the case-study of a farm; to Bernard Johnson of Coubaril for his information on the case-study of a manufacturing and service industry, and to Hubert and Marie Dennehy of Cul-de-Sac for information on the banana industry, and also their generous hospitality.

So many other people have helped in smaller ways that it is not possible to record them all here. My sincere thanks are due to all those with whom I worked in Saint Lucia, and especially Mrs Marjorie Braithwaite of the Teachers' College, for generally accommodating my interest in the island and its development.

The author and publishers wish to acknowledge the following photographic sources:
Anne Bolt
J. Allan Cash
Eastern Caribbean Commission
Geest Holdings Ltd.
Popperfoto
Public Relations and Information Department, St. Lucia
Dr Claudius Thomas, Eastern Caribbean Commission
West India Committee

Preface

In many countries of the world the name Social Studies is appearing on school curricula and timetables. This new title, together with others like Environmental Studies and Humanities is tending to replace names like Geography and History. The object of this is to allow certain other subjects like Sociology, Economics and Politics to play their part in the general education of school children.

This particular book has been written as a contribution towards the stock of material available to teachers of Social Studies in Caribbean schools, and especially in Saint Lucia. It is not so much a text, as a background book which together with other workbooks and texts in this series should enable the teacher and student to follow a course of home-based Social Studies. While intended mainly for the Junior Secondary Schools it is hoped the other groups in primary, secondary and further education establishments may find it useful.

The author worked in Saint Lucia during 1972 and 1973 and became acutely aware of the need for extra teaching material in this field. The book is offered in the hope that those teachers and colleagues with whom he worked will find it a useful first step in the development of resources to assist their efforts.

To Leyton Thomas and George Delmede
servants of Saint Lucian education

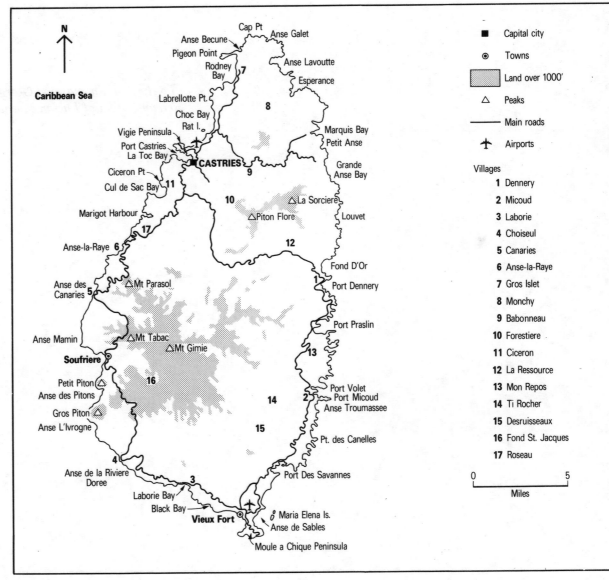

Fig. 1 St Lucia: places

Saint Lucians

Social Studies is to do with people and how they live in any particular place. This book is about the people of Saint Lucia. The people are Saint Lucia.

Saint Lucians share many common characteristics with other people all over the world. And yet this little nation is unique, as all true nations are. So in the following pages we shall find that Saint Lucia has its share of the world pro-blems of population increase and food supply. But on this island the reasons for such problems are, in detail, different from other places. The solutions likewise will have to be different.

A nation's greatest resource is its people. Let us ask: 'Where are the people?' 'Who are the people?' 'What do they do?' 'How and why are the numbers changing?'

Where are the people?

Most atlases of the Caribbean area include maps of Saint Lucia rather like the ones below. Anybody who knows Saint Lucia will notice immediately that the first map is not a true picture of where the people really are. The map is out of date, as most maps are! It gives a fair idea of where the Saint Lucians were living in about 1930.

If we make an intelligent guess at where the people are now, we might agree on something like the second map.

Both maps attempt to show **distribution of population.** Obviously this is affected by the mountains and the rivers, yet it changes all the time. So we will begin our study of Saint Lucia with the people rather than the land.

Why has the pattern changed?

The distribution of population depends on the way of life of the people, and the nature of the land. Since 1930 there have been great changes in the way of life, or life-style, of many Saint Lucians.

However, the population pattern of 1930 was based upon the economic history of the island over hundreds of years. The colonisation of Saint Lucia had traditionally been from the south, and the south-west in particular has always been sheltered from the north-east trade winds. The French had their capital at Soufriere, and then, after emancipation, freed slaves set up many small farms in the area. Over the years, land was subdivided and the number of farms increased.

Fig. 2   Where the people are now

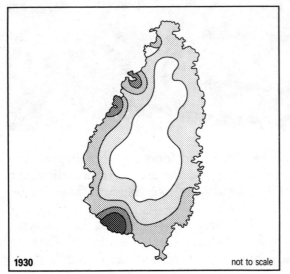

1930   not to scale

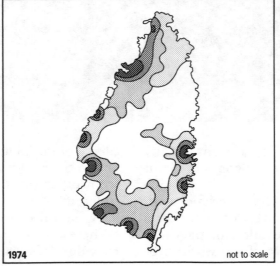

1974   not to scale

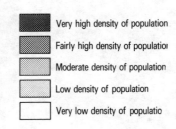

■ Very high density of population

▨ Fairly high density of population

▦ Moderate density of population

▫ Low density of population

□ Very low density of population

Fort. The trade of the village ports has almost vanished.

3 Statehood, officially achieved in 1967, has led to the development of jobs for many more Saint Lucians in the administration of their nation. However, most of these jobs are in Castries, the modern capital.

4 Population growth has become more rapid, mainly due to improved medical services. These tend to help more babies to survive, and old people to live longer.

This has created economic growth and a whole range of new jobs have become available. Tourism is one growth area, retailing and finance are others. People no longer have to be peasant farmers. Young people especially are attracted by the opportunities available in Castries.

Throughout the colonial period, under the French or British, Saint Lucians had little opportunity to get non-agricultural jobs. Most of the administration was from Paris or London. Sugar was the main export, and was shipped from a number of different ports. As links between villages were best by sea, each one was the focus of its own little area or hinterland. There was no motor transport so journeys by road had to be short. Castries could offer nothing that most of the villages could not provide themselves.

In colonial times medical services were poor and the death rate high. Population increase was steady but slow. Most people were peasant farmers, and lived a rural life.

Many of these things have changed since 1930, but four need special mention.

1 Motor transport. The development of cars, lorries and buses has brought all the villages and out-districts within easy reach of Castries. Therefore Castries, as a market, has gained at the expense of the villages.

2 Bananas have taken over as the main export and are shipped mainly from Castries, with some from Vieux

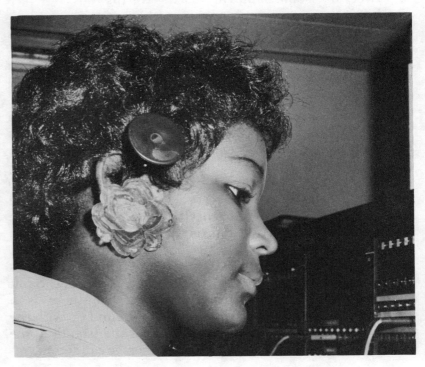

The pattern of population is therefore one of urbanisation, a movement away from the land. Not only has there been a shift of focus from south-west to north-west, but the **density of population** has increased. This means the average number of people per unit area of land, for example, 500 per square mile, is greater.

Why are the numbers changing?

The number of people in any place doesn't stay the same for long, and Saint Lucia is no exception.

   Population change is caused by three simple things:
1   Every day some babies are born — births
2   Every day some people die — deaths
3   Every day some people leave the island, and others come in — migration.

In most countries this change is usually an increase. This is because there are more births than deaths each year, and fewer opportunities to gain employment overseas.

   Accurate details of population change are obtained in two ways. First, there is the recording of all births, deaths and migrations, which is called vital registration. Secondly there is a general survey made about once every ten years, called a census. Saint Lucia had its latest census in 1970 when 101 064 people were recorded. The 1960 census had shown 86 108. According to the *West Indies Chronicle* in June 1975, the population was approximately 140 000.

Births   The crude birth rate means the number of babies born for every 1000 people in any one year. The Saint Lucian rate is comparatively high because there is a tradition of having large families. This is bound up with the economic, social and religious history of the island. Some islands have similar traditions while others are different.

Approx. crude birth rates of some Caribbean States 1970

Haiti	45	(Highest in the Caribbean)
Dominican Republic	41	
SAINT LUCIA	39	
Trinidad & Tobago	35	
Martinique	34	
Puerto Rico	30	
Barbados	27	(Lowest in the Caribbean)

Deaths   The crude death rate means the number of people dying for every 1000 people in any one year.

Approx. crude death rates of some Caribbean States 1970

Haiti	27	(Highest in the Caribbean)
Dominican Republic	14	
Virgin Islands	9	
SAINT LUCIA	8	
Trinidad & Tobago	7	
Netherlands Antilles	5	(Lowest in the Caribbean)

Migration   Migration may be internal, from Choiseul to Castries, or international, from Saint Lucia to England. It is the international migration that affects population change. People who leave are called emigrants, those who come in are immigrants. Traditionally young men have left the island for a few years to seek well paid employment overseas. In recent years such opportunities have been strictly limited by immigration regulations elsewhere.

   Internal migration is mainly a movement towards Castries, but most villages are gaining at the expense of rural areas. This is called rural depopulation. The rate of population change may be shown by the following table.

Growth of population in Saint Lucia 1960-1975

Year	Total	Source
1960	86 108	1960 Census
1970	101 064	1970 Census
1975	140 000 (estimated)	1975 (May/June edition *West Indies Chronicle*)

Who are the Saint Lucians?

People are different. We tend to fall into groups. For example there are different sexes, races, ages, religions, families and occupations. Such groups make up the **population structure.** The most obvious facts about Saint Lucians are that most are black and young.

Races All three major human racial groups are represented on this island. Their relative proportions tend to reflect its economic history.

The earliest inhabitants were Amerindians of mongoloid race, from South America. Very few Saint Lucians are descended from these people. Next came the Europeans, mainly French and British, of caucasoid race. These were even fewer than the Amerindians, but are still important today. The Europeans brought African slaves to Saint Lucia, who were, of course, negroid in race. The Africans were far more numerous than the others, and, despite racial mixing, pure negro race is the most common type today.

In about 1880, the British introduced a new labour force to the plantations of Saint Lucia. They were called indentured labourers, and came mainly from India. They are often called 'East Indians' but they do not, of course, come from the East Indies. About 4000 came, and in turn have mixed with those of African and European origin. Finally,

in Castries there is a significant community of Syrians. These people come from the Levant coast of the Eastern Mediterranean, and similar communities are found in most of the world's seaports. Both Indians and Syrians are of caucasoid race. People of mixed race today form about a quarter of the Saint Lucian people. They are sometimes called Creole.

Occupations People may also be classified according to the type of work they do. Clearly agriculture is still the

Fig. 3 St Lucia: diagram of occupational structure c. 1970

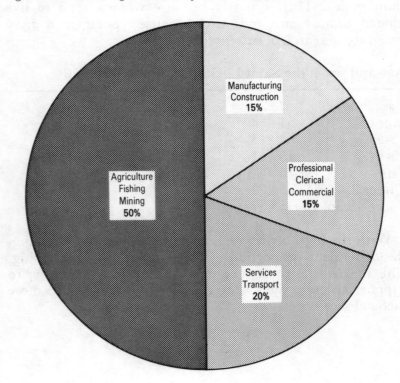

5

dominant area of work in Saint Lucia. Diagrams like this do not include the enormous amount of unpaid farm work done by women, children and old people. Yet the diagram also shows that many other types of employment now exist. We may say that the employment structure has been diversified.

Age and sex   It has already been mentioned that most Saint Lucians are very young. In fact about half are aged fifteen or less. This is because of the tradition of early childbirth and large families.

The following table shows that there are more females than males. This is partly because women tend to live longer than men, and also because young men may emigrate in search of better work.

Age and sex of the Saint Lucian population in 1960

Age Group	Male	Female	Total
over 65	1 454	2 668	4 122
45-65	4 813	5 948	10 761
15-44	15 161	17 957	33 218
2-14	16 043	15 667	31 397
Under 2	3 222	3 175	6 397

Source Saint Lucia Census 1960

With the high birth rate, and more old people surviving, the dependent proportion of the population is increasing. This means that those who do work and pay taxes have to support an ever increasing number of schools, and more medical services.

Religion   Most Saint Lucians are Roman Catholic, but several other churches are represented on the island. The Catholic Church is dominant because of its firm foundation in the growing communities of the island during the French colonial period. During the present century Nonconformist groups, mainly from North America, have greatly increased their influence. The Seventh Day Adventist Church is the best example of this in Saint Lucia.

Family structure   The family, in various shapes and sizes, is the foundation of society all over the world. We may think of the family in different ways; perhaps just as mother, father and children. This is called the nuclear family. Then there are grandparents, uncles, aunts and cousins making up the extended family. Or we might think of the home, the people and the building, living together to form the household.

To be related to somebody through blood or marriage is called kinship. Saint Lucians have a strong sense of kinship. Inevitably there is a great deal of inter-relationship within a small island community like this.

Key words

population	births	depopulation
resources	deaths	family
change	migration	race
urbanisation	census	employment
sex	religion	kinship

Communities

People tend to live in groups. Very few people live entirely on their own. Man is a co-operative animal in that the task of survival is shared. We need each other. We are inter-dependent at all levels: local, national and international. We form families and nations, but between these two extremes we live our everyday lives in communities.

In simple societies one small community hangs together.

In advanced societies each person belongs to many communities at the same time: the home, the church, the football team, the school.

Formerly, communities were places on the map: estates, villages, a street in the town. Each settlement was a real community. Everybody belonged there. Now Saint Lucia is becoming one national community focused on Castries.

Schools and communities

A school is an example of a community within a community. A group of people working together in one place for a particular purpose.

In 'primitive' societies schools did not exist at all. Teaching was done by parents and older members of the tribe or clan. In more advanced communities like Saint Lucia, it is not possible for parents to teach their children all that they need to know in the modern world. As schools develop, and especially secondary schools, they help to change the structure of society.

Primary schools Most infant, primary and combined schools in Saint Lucia serve a distinct rural community. For generations these schools have been a part of local life.

Secondary schools Until 1970 there was little opportunity for most Saint Lucian children to attend secondary school. Only one of the secondary schools was outside Castries. In other words, secondary education was not part of local life, it was remote and urban.

Since 1970 eight Junior Secondary Schools have been opened. Most of these are in the out-districts, not exactly local, but certainly within reach of the local community. By studying any of these schools it is possible to outline the main needs of any type of community. These include an area of land; a set of buildings; leadership; responsibility and co-operation; some regulations; being part of both local and national communities at the same time.

A school is a particular type of community linked with the needs of other communities. Its students belong to many other communities and groups at the same time. School is for learning. We are all interdependent. Our

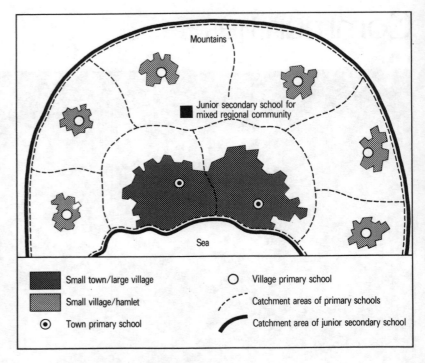

Fig. 4 Comparison of school catchment areas

backgrounds are rural and urban, local, national and international. Old Saint Lucia was colonial and local. New Saint Lucia is independent and national, one community. A nationwide pattern of secondary schools is a necessary part of community development in Saint Lucia.

Local community

Schools provide both an individual and a national service, but people have other needs that are more immediate than formal education. Out of the fulfilment of these needs grew the idea of local community in Saint Lucia. Two needs are very special: the need for security, and the need for basic materials, like water and soil.

Security   Everybody needs security, which means they need to be, and feel, safe. Here are some obvious examples
1   Young children are completely dependent, they need others to survive.
2   Everybody needs protection against the weather and some other people.
3   Most people need mental or spiritual security against unknown things which they do not understand.
4   People need economic security, which means the ability to obtain sufficient food.

Basic needs like these are met by the formation of local communities. In the days of slavery, the plantation community was part of an international economic system; it was not of the island. When slaves escaped or were freed, they set up home in the hills and were faced with these basic needs. The answers they found were similar to those of other peoples. In order to meet the needs of dependence and protection they developed the family and the home. Hitherto slaves had not been allowed to live in family groups.

Economy and basic materials   Food is necessary for survival. The freed slaves of Saint Lucia had a very limited choice, a choice between farming and fishing. Families grouped together in places where they could best carry out these activities. The best lands were, however, still used for plantations. So communities began to develop on the steep ridges between plantation valleys, and also at some places on the coast. Some people decided to continue working on the plantations and communities developed there also.

The people and buildings of any place are called settlements. Families build homes, religious groups build churches. Community resources develop, like a well or a post box. Each community in old Saint Lucia also developed its own system of **local responsibilities.** Respected individuals would be entrusted with certain tasks.

Gradually, churches and primary schools were established in the larger communities. Local councils were formed and were able to represent the people in seeking better basic services. Local markets were built in some places, shops were opened, communal water supplies and eventually even electricity arrived.

Some communities grew, others did not. Some even declined. Certain places became known as the focus of their district. Nowadays we call them the villages of Saint Lucia.

Canaries: a St Lucian village

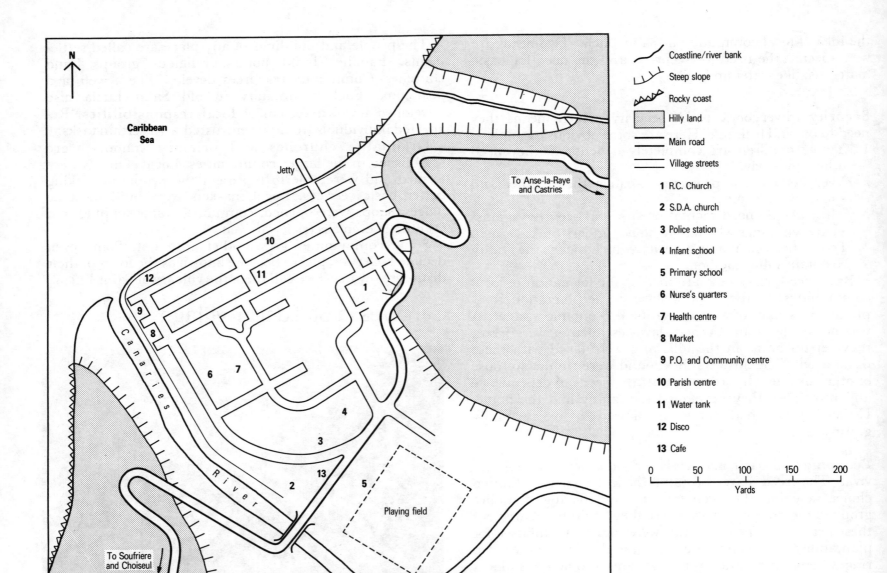

N

Caribbean
Sea

Jetty

To Anse-la-Raye
and Castries

C a n a r i e s

R i v e r

10

12

11

9

8

1

6   7

4

3

13

2   5

Playing field

To Soufriere
and Choiseul

Coastline/river bank

Steep slope

Rocky coast

Hilly land

Main road

Village streets

1 R.C. Church

2 S.D.A. church

3 Police station

4 Infant school

5 Primary school

6 Nurse's quarters

7 Health centre

8 Market

9 P.O. and Community centre

10 Parish centre

11 Water tank

12 Disco

13 Cafe

0   50   100   150   200

Yards

Fig. 5 Plan of central Canaries

Every community and village is different in detail, but they all have similar basic needs and services. A study of one village in detail will serve to illustrate the basic structure of a Saint Lucian community outside Castries.

The map opposite shows clearly that the structure of a Saint Lucian village is no longer rural. Canaries is part of the changing network of a developing nation.

Tasks which would once have been carried out informally have become institutionalised. Special buildings have been erected where different community services are carried out. These are on the map; buildings like the post office, police station, health centre and community centre are fairly recent improvements on the old services. Likewise, old Canaries would not have had a disco or a café. Other services are more traditional, and long standing, like the school, church and market.

Looking at the village today we can see something of its history. The old centre of the village seems to be laid out in

a careful plan of parallel streets. It was most probably a plantation settlement aligned to the main plantation buildings as indicated in the century-old photograph on page 11. The village existed for the housing of estate workers, and the export of the products. Obviously the sea and the river were important in the siting of the village.

Since about 1950 the estate and the sea have become less important to Canaries than the new road from Castries, and the village has grown towards the road and the bridge. Canaries is now the centre of a wider community, from Anse-la-Verdure to Belvedere, based on the road.

Parts of the old estate area have been sold, and the village has been able to expand across the road and up the hillsides. Like most Saint Lucian villages it needs to grow. Some people wish to leave the land and join the village. Also, new generations need houses.

Growth has not always been the rule. Like other villages, Canaries lost many of its young men and families between about 1945 and 1965. They emigrated to seek employment in Curaçao, North America and mainly in England. Even so, family and community ties remain strong. Regular remittances of money are sent back home and visits are made, to maintain close contact.

Canaries is now growing again. It has its own character. Families there have known each other for generations. There is a sense of community, a sense of belonging. There are many problems, but the basic needs of security and services are provided for.

Communities and settlement

In a general sense Canaries is one community, but really it consists of many communities. For example there are the fishermen, the shopkeepers and the schoolchildren. One

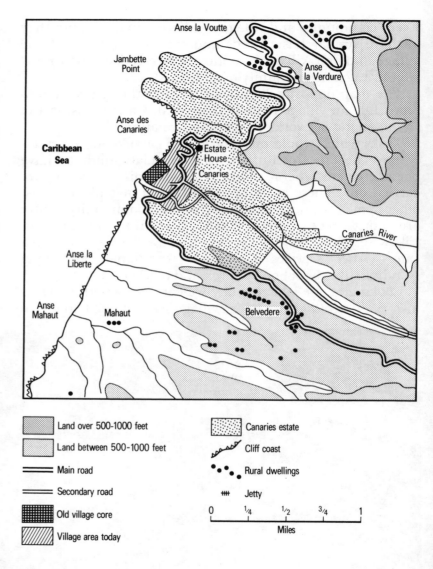

Land over 500-1000 feet

Land between 500-1000 feet

Main road

Secondary road

Old village core

Village area today

Canaries estate

Cliff coast

Rural dwellings

Jetty

Fig. 6 Site of Canaries

person may belong to many human groups at the same time, for example, a football team and a church. Communities overlap each other.

The study of human groups and how they interact is called **Sociology.** Later in the book we shall be looking at different groups and occupations, but here we begin with the most obvious sign of a community, the settlement.

A settlement is a place where people live and/or work. For example it could be an individual house on a farm or estate, or it could be a village, a town, or even an isolated hotel. Because a settlement is a place on the earth's surface, its study forms part of the subject known as **Geography.** There are two main types of settlement, rural and urban.

Rural settlements   These are places where most of the people work on the land. They are small because the land is needed for food production. Rural settlements may be either dispersed (scattered), or nucleated (clustered). For example, Anse-la-Verdure is a dispersed settlement. There does not seem to be an obvious centre or focus.

By contrast, this part of Roseau is a nucleated settlement. Services are located here for the people round about: a school, a church, a community centre, a shop. Obviously nucleated settlements tend to be larger, and are more important community centres. Small fishing villages are a type of rural settlement.

The patterns of rural settlement show three main types of location.
1   Coastal, usually riverside sites (e.g. Gros Islet).
2   Inland valley sites (e.g. Roseau, Bexon).
3   Hilltop sites (e.g. Mon Repos, Forestiere).

Urban settlements   These are places where most people do not work on the land. They are usually called towns,

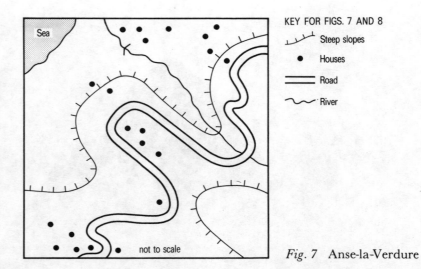

KEY FOR FIGS. 7 AND 8

⊥⊤ Steep slopes
● Houses
══ Road
∿ River

Fig. 7 Anse-la-Verdure

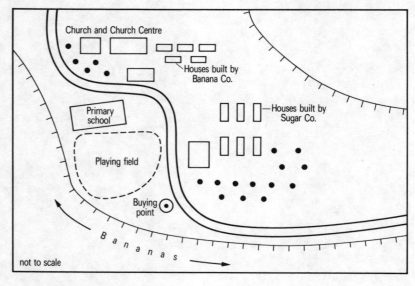

Fig. 8   Part of Roseau

13

and are places where goods are exchanged and services available. For example, a town would normally have banks, a variety of specialist shops, and possibly agencies and other businesses. They would also tend to have cafés and places of entertainment. Most of the large Saint Lucian villages have a mixture of urban and rural ways of life, but there are only two towns, Soufriere which is shown opposite and Vieux Fort, plus the city of Castries.

On this map the settlements are shown according to their status. The status of something is its importance or rank in relation to other things of the same type. The map is based on the results of a survey made by students of Saint Lucia Teachers' College in 1973, and is based on the services available in each place at that time.

Settlement patterns are always changing. In Saint Lucia, as in many parts of the world, people are leaving the land to live in towns. This process is called **urbanisation.** The growth of Castries is a good example of this. Compare this map with those on page 2.

Castries Towns and cities are centres of exchange. This is because they are route-centres, places on which networks of transport and communication converge.

Castries is a special type of route-centre in two ways, as a port and a capital city. Central Castries, and its harbour, is the focus of the nation. Here, shipping routes from many parts of the world meet the road routes coming in from all parts of the island.

The growth of Castries has taken place in three stages.
1 Foundation by the French. In 1763 Castries became the new centre of administration in Saint Lucia. It replaced Vigie which, just a very short time before had replaced Soufriere as the capital.

At that time, competition for the island was fierce

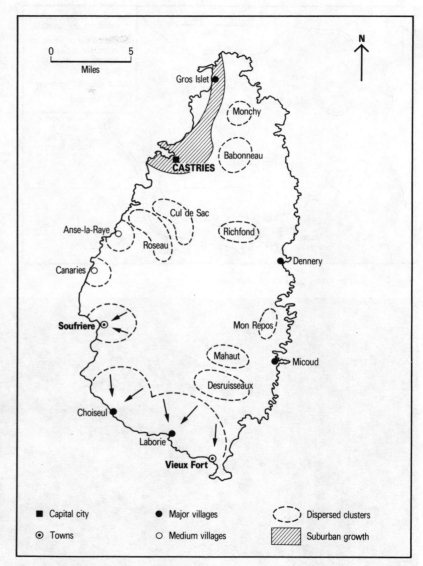

Fig. 9 St Lucia: status of settlements

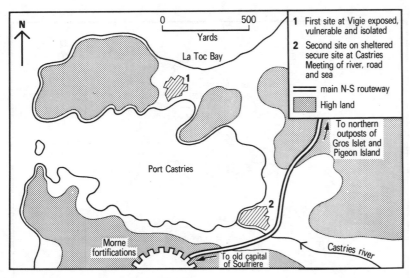

Fig. 10 Early sites on Port Castries

between Britain and France. Castries, at the head of a long inlet, and beneath the fortress on Morne Fortune was a good military site. It was built on flat land reclaimed from the harbour and bounded inland by the Chaussée Road.

2   Bunkering Station. After 1830 there was no further competition for the island. For fifty years Castries remained the same. Then steam replaced sail, and the British Navy chose Castries as its main coaling station in the West Indies. Merchants competed to supply ships and sailors with all kinds of goods, and thousands of rural people came to Castries seeking work. The merchants built great houses on the hillsides, and the others clustered in crowded new suburbs around the edge of the old town.

3   State Capital. Soon after 1920 the coal boom ended. All over the world, trading declined. Castries became quiet and local once more. Then in 1958 the whole West Indies became independent. Saint Lucians were running their

own country and many new jobs in the civil service were needed. Castries became a real capital. New features of the economy became centred on Castries; the export of bananas and the development of tourism.

Once again a great wave of migration built up, and Castries now comprises about 40% of the population. In addition to further overcrowding of the older suburbs, new wealthier areas of housing have spread. Castries is now Greater Castries, a conurbation from Cap to Cul de Sac, and Vigie to Babonneau. In reality Saint Lucia is a City State. There is no rival to Castries now.

Castries is many communities, and it also has many distinct areas or zones. The old town is still the core or focus of government, commerce and industry. Around this core are many suburbs, some old, some new. Some are purely housing, others are part of the tourist trade. Urbanisation on this scale creates problems for community life. Suburbs are usually places of private living, not communities. Social classes may become more separated, and 'communities' a complicated network of contacts rather than a place on the map.

Key words

interdependent	settlement	council
community	school	market
services	local	remittances
route-centre	national	economy
urban	suburb	international
materials	rural	village

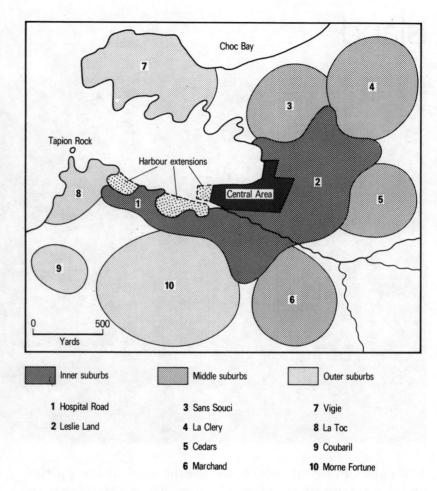

Fig. 11 Suburban structure of Castries

Inner suburbs **Middle suburbs** **Outer suburbs**

1 Hospital Road	**3** Sans Souci	**7** Vigie
2 Leslie Land	**4** La Clery	**8** La Toc
	5 Cedars	**9** Coubaril
	6 Marchand	**10** Morne Fortune

Island

The communities in which we live or work make up the **social environment.** Most social activities are related to particular places. For example homes, schools, churches and cricket matches all have a definite location.

Our lives are also closely linked to the **physical environment,** that is, the land, air and water around us. About two thirds of the surface of our planet Earth is water. Most of the remaining third, the land, is too cold, too high or too dry to support human life. So we must be very careful not to destroy the small amount of good land available.

We depend on the natural earth for every material thing, our food, our wood, our metal, our textiles, our power supply. Clean air and water are essential too. What does the physical environment of Saint Lucia consist of?

Weather and climate

People don't just live on the land surface, they live at the same time in the atmosphere. Without the atmosphere we should all die. We need the oxygen and the rainwater.

Weather, means what is 'going-on' in the air at any particular time. It may be raining, or cold, or windy, or the pressure may be low. The study of weather is **Meteorology.**

Climate means the average weather at a place, year after year. Saint Lucia has a tropical maritime, (warm and moist), climate. The study of climate is **Climatology.**

Temperature The most important element is the supply of light and heat from the sun. This supply is called insolation. The measurement of heat is called the temperature, usually given in degrees (fahrenheit or centigrade). Below are the average monthly temperatures for Castries, compared with London.

Precipitation This means the supply of moisture from the air to the ground, usually in the form of rainfall. Rainfall is measured in inches or in millimetres. Below are the average monthly rainfall figures for Castries compared with London.

Together the heat and moisture serve to develop the soil, and keep all plants and animals alive, at least for a certain period of time which we call their life-span.

Pressure Atmospheric pressure is the force of the air pressing down and in upon us. If you go high up, on the Morne or Barre de L'Isle, the pressure is less and therefore the temperature is lower. On a large scale, differences of

Mean monthly temperatures for Castries and London

		Jan	Feb	March	April	May	June	July	Aug	Sept	Oct	Nov	Dec
Castries	Fahrenheit	79	79	80	80	81	81	81	82	82	82	81	80
	Centigrade	26	26	27	27	27	27	27	28	28	28	28	27
London	Fahrenheit	41	41	43	47	55	59	63	62	57	51	44	41
	Centigrade	5	5	6	8	13	15	17	17	14	11	7	5

Mean monthly precipitation for Castries and London

		Jan	Feb	March	April	May	June	July	Aug	Sept	Oct	Nov	Dec
Castries	inches	5.4	4.0	3.6	3.5	5.7	9.2	9.3	10.2	9.3	9.6	8.9	7.4
	mm	137.2	101.6	91.4	88.9	144.8	233.7	236.2	259.1	236.2	243.8	226.1	187.9
London	inches	1.8	1.5	1.7	1.5	1.7	2.1	2.2	2.2	1.9	2.7	2.2	2.3
	mm	45.7	38.1	43.2	38.1	43.2	53.3	55.9	55.9	48.3	68.6	55.9	58.4

pressure cause movements of air called winds. Saint Lucia happens to lie in the path of the north-east trade winds.

These winds not only supply moist air to provide rain, they also give a cooling breeze to relieve the heat of the sun. So the sun, the rain and the trade wind breeze combine to give Saint Lucia one of the most pleasant climates in the world. Another advantage of having plenty of warmth and moisture is that plant growth is possible all year round. With care, local food should always be available.

The following diagram illustrates the idea of the water-cycle with reference to Saint Lucia.

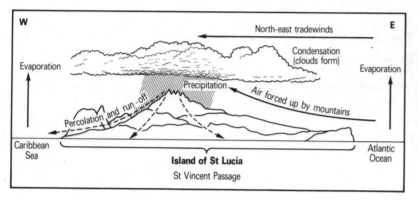

Fig. 12 Water-cycle applied to St Lucia

Vegetation

Natural vegetation Because of the climate, the main example of natural vegetation here is tropical rain forest. This dense tree growth is possible wherever rainfall is in regular supply, but in the northern and southern limits of the island it is not. As a result, conditions in these areas are hot and dry, creating zones of semi-desert where xerophytic plants like cacti are found.

Modified vegetation Over most of the island man has changed the vegetation through various forms of agriculture. Much of the forest of the plateau areas is mixed forest, including fruit trees introduced from other parts of the world (orange, lime, breadfruit).

Where the forest has been cut down, and the land abandoned, then secondary vegetation replaces the agricultural land. In wet areas this takes the form of secondary forest, lower and thicker than the original forest. In drier areas a type of scrub or thorny bush takes over, and is now widespread.

How was the island formed?

Most Caribbean islands are either (a) coral islands or (b) volcanic islands. Saint Lucia is a volcanic island with some coral formations around the edge. Both coral and volcanic types are of recent geological origin, and have developed over the past twenty-five million years.

The present island of Saint Lucia is all that remains of a great volcano once centred on what we now call Mount Gimie. All that we can now see is just the very top of an enormous mountain that sits deep down on the ocean floor. It took millions of years for the island of Saint Lucia to form, in a number of stages.

1 A series of small volcanoes formed an island of what is now the area north of Castries.
2 Gradually this land was eroded and lowered as the volcanoes died out.
3 Then a new larger volcanic island developed to the south, centred on Mount Gimie. This became joined to the lower northern part.
4 Just as the whole area was being eroded again another series of volcanoes became active. This eruption was

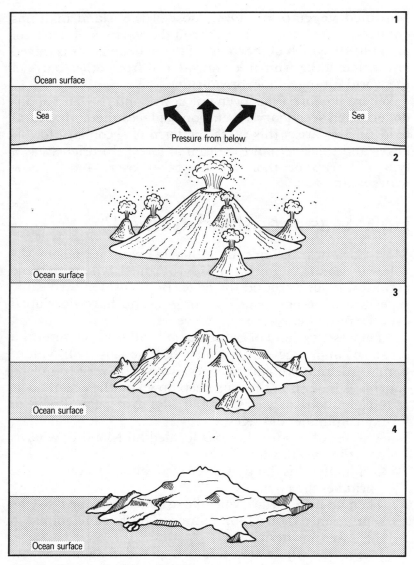

Fig. 13 How St Lucia was formed

centred on Soufriere Bay, and half of the new land sank into the Caribbean Sea. The other half remains as the Pitons and Sulphur Springs.

In some parts of the world it is still possible to see new volcanoes rising out of the sea from time to time.

Rocks

The earth, our planet, is made up of three main types of rock. The study of rocks is **Geology.**

1 Deep inside the earth the heat is tremendous, and there are areas of magma. These are masses of liquids and gases all moving about. When a volcano erupts, the magma flows out onto the surface as lava. As the lava cools it becomes solid rock. Any rock which forms from magma is called an igneous rock. Most of Saint Lucia was formed from this type of rock.

2 If any rock becomes very hot, or is put under pressure it will change. Rocks which have changed their nature are called metamorphic rocks. They are often found to contain useful mineral deposits, especially the ones of metals, like iron.

3 A third type of rock may be formed by the accumulation of tiny rock particles under water, or on a desert surface. These are called sedimentary rocks.

Landscape

Landscape means scenery. Hills and valleys, deserts and forests, are some of the features of the natural landscape. But we must not forget the banana plantations, the citrus plantations and the harbour extensions; all of which are part of the human landscape. Where big cities like Castries completely cover thousands of acres of land with concrete, we call the effect the townscape. Put all these things together and we have the environment.

It has been said that the land of Saint Lucia has a certain structure, depending on its rocks. As soon as the island was formed it began to be changed in shape by two processes of nature, called erosion and deposition. Erosion means 'wearing-away' and deposition means 'building-up'.

Erosion and deposition are caused by things called agents, like the wind, the rivers and the sea. Because of this, the shape of the land surface is slowly changing all the time. Man can be an agent of erosion, for example by farming steep slopes and exposing the soil to heavy rain. Or he may be an agent of deposition, as with the construction of the causeway to Pigeon Point near Gros Islet.

In general there are three main types of land surface: (a) mountains and hills, (b) plateaux, (c) lowlands. To these

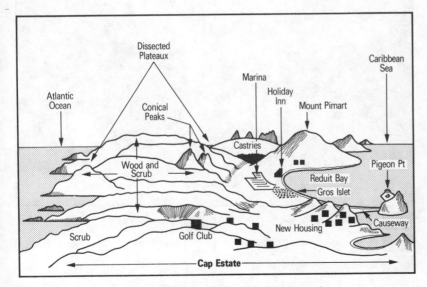

Fig. 14 Field sketch from Pointe du Cap looking south

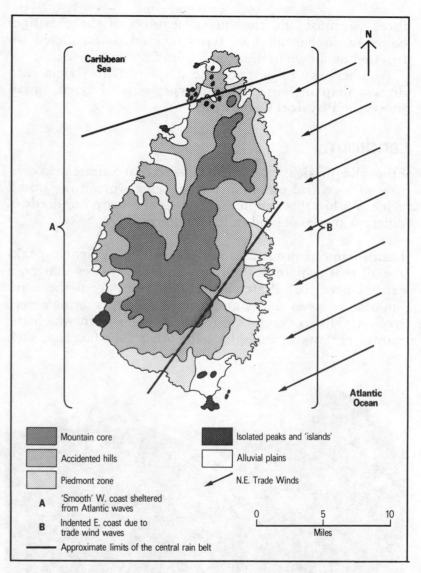

Fig. 15 St Lucia: physical zones

three we must add the special features of the coastline. Saint Lucia has all four types of land, often mixed up together as shown in the sketch from Cap on page 23.

Another way to describe the shape of Saint Lucia is to draw a map showing different types of land. Such a map shows the **Physical Geography.**

Landforms

From the physical map it can be seen that Saint Lucia consists of a central mountain range being worn away on all sides. Within this simple framework there are hundreds of different shapes of land called landforms.

Landforms of mountainous places (a) Volcanic peaks are all that remain of the many small volcanoes that once existed here. (b) Watershed ridges come from the combination of steep slopes and heavy rainfall, creating severe erosion. This is very obvious along ridges from which the rainwater flows to one side or the other. The fact that such

a) Contour map (vertical interval 50 feet)

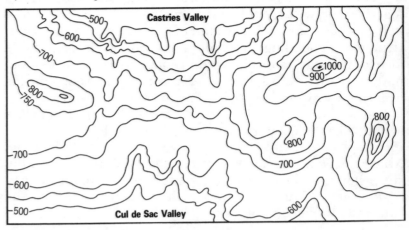

b) Selected altitudinal zones

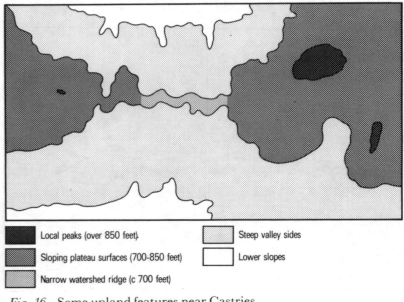

Local peaks (over 850 feet).

Sloping plateau surfaces (700-850 feet)

Narrow watershed ridge (c 700 feet)

Steep valley sides

Lower slopes

Fig. 16 Some upland features near Castries

24

ridges tend, in Saint Lucia, to be popular zones of settlement has the effect of increasing the erosion. This is due to cutting down the forest cover and exposing the soil.

The most important thing about mountain areas is not how high they are but how steep the slopes are. Steepness of slope can be shown on a map by using contour lines. When the lines are close together then slopes are steep.

Landforms of plateau areas   Plateaux are areas of high flat land. In fact erosion causes most plateaux to be sloping and also to be cut-up, into blocks, by river valleys. Plateaux like this are called dissected plateaux, and are found all round the edges of the central mountainous area of Saint Lucia.

River valleys   When rainwater falls on the land it flows downhill under the pull of gravity. Over thousands of years this water finds the easiest route to the sea and wears it into a groove, or valley. Valleys are found in mountain, plateau and lowland areas.

1 Young river valleys are steep-sided with narrow bottoms. They are formed where erosion is more active than deposition.
2 Older river valleys may be steep-sided but have wide flat floors. Here, deposition of sediment is important, especially filling in areas where coastal inlets once existed. Cul-de-Sac and Roseau valleys are good examples of this type. The sediment in these valleys may be over a hundred feet thick.

Lowlands   The older river valleys are one type of lowland. As they get near to the sea some of them develop into deltaic lowlands. All round the coast of Saint Lucia, sediment has been building up for millions of years. In some areas very extensive flat lands or plains have been formed. When the rivers are in flood they spread thin layers of sand and silt over the lowland, thus gradually building it up. Sometimes they join together hills that once were islands, as in the Vigie area. This process has been carried on by man in the form of the Castries harbour extensions.

Coastal landforms   Deltaic lowlands are both land and coastal features at the same time. They are features of coastal deposition. In other places where the land plunges steeply into the sea, the coastal erosion is more important. This is particularly common on Saint Lucia's Atlantic

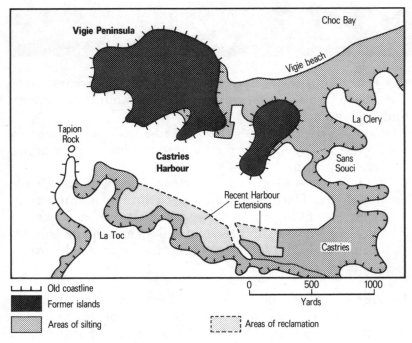

Fig. 17   Vigie peninsula

coast where the trade winds built up massive waves to batter the cliffs. Coastlines of erosion tend to be very indented, while coastlines of deposition are smoother in outline.

Soil

The soil is the most vital part of the environment. It is formed by a dynamic relationship between the atmosphere, the rocks and the vegetation. Together with these factors, and with animal life, the soil makes up the ecosystem of any particular land area.

Soil is made up of mineral particles from the rocks, the remains of dead plants and animals, called humus, plus air and water. Usually, soils derived from volcanic rocks are rich in mineral content, and are fertile. This is the case over much of Saint Lucia. Soils formed on limestone/coral or sand may be thin and sterile.

Soil erosion   Soils are delicate. They can easily be destroyed. Every time it rains some of the soil is washed

26

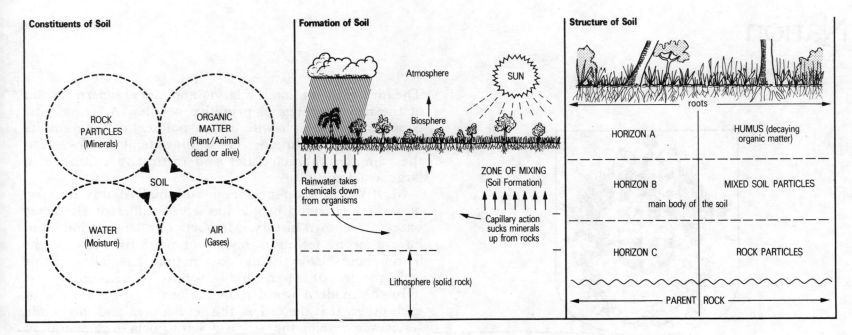

Constituents of Soil

ROCK PARTICLES (Minerals)

ORGANIC MATTER (Plant/Animal dead or alive)

SOIL

WATER (Moisture)

AIR (Gases)

Formation of Soil

Atmosphere

SUN

Biosphere

Rainwater takes chemicals down from organisms

ZONE OF MIXING (Soil Formation)

Capillary action sucks minerals up from rocks

Lithosphere (solid rock)

Structure of Soil

roots

HORIZON A — HUMUS (decaying organic matter)

HORIZON B — MIXED SOIL PARTICLES

main body of the soil

HORIZON C — ROCK PARTICLES

PARENT ROCK

Fig. 18 The world of the soil

down to lower ground, and some ends up under the sea. Because of this, valley bottoms often have deep soils, while ridges and hillsides have shallow soils.

We have already seen that for historical reasons the ridges of Saint Lucia are covered with small farms and settlements. This tends to increase erosion, and supplies soil to the estates in the valley bottoms. Where larger farms exist it may be possible to check soil erosion by introducing contour ploughing. In this, and other careful ways, the soil of Saint Lucia may be conserved for the future.

Key words

atmosphere	precipitation	coral
oxygen	pressure	volcanic
plateau	rainfall	wind
rock	valley	weather
trade winds	landscape	ridge
climate	water-cycle	environment
lowland	tropical	vegetation
erosion	plain	element
rain-forest	deposition	cliffs
insolation	semi-desert	landform
soil	temperature	mixed-forest
contours	ecosystem	humus
fertile	sterile	conservation

27

Nation

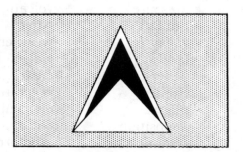

The idea of the nation is a larger and more modern version of the tribal feelings of primitive peoples. A nation is a community of people with a clear political identity and its own political institutions. These things tend to grow out of the common characteristics and distinctive culture of a large social group.

Most West Indian societies live on small islands. Because of this, each island people has a very different character based on its own history. Members of a true nation often have a strong feeling of loyalty or patriotism towards it. Throughout history many older nations have allowed this feeling to get out of hand and wars have been caused by it.

In the modern world most nations try to coexist with each other. It is the task of the government and the public services to develop the nation of Saint Lucia in co-operation with other nations, in the Caribbean and beyond.

The making of the nation

National style or character is something that usually develops before political independence. With independence, or at least internal self-government, steps are taken to develop national institutions. This is where a national community develops a full political structure, and becomes a nation-state. Despite her small size, there is no doubt that the island community of Saint Lucia is a genuine nation.

Political history is the study of the evolution of systems of administration in any particular place. For Saint Lucia we can trace that development through several stages. First, the tribal societies of the Arawaks and Caribs, then the colonial systems of the French and the British and independence within the Federation of the West Indies. All of these things came before Associated Statehood, which began in 1967. Full Independence is planned for 1976.

Amerindians

Today the Amerindian contribution to Saint Lucia is nil, but we should look briefly at those first Saint Lucians who called the island 'Iouanalao'.

Three groups of Amerindians (American Indians) occupied the island in turn (a) the Ciboney, (b) the Arawaks, and (c) the Caribs.

Although the Arawaks were more peaceful and civilised than the Caribs, both had a tribal organisation. The local leaders had almost complete control of affairs. Much of their culture shows some connection with the great civilisations of Central America, the Mayas and the Aztecs. The many archaeological remains on the island help to give some impression of a pre-colonial nationhood.

French Saint Lucia

There are many different ideas as to when Saint Lucia was 'discovered' by Europeans. It could not have been before the first voyage of Christopher Columbus in 1492. In any case, the Spanish did not colonise the island, and it is likely that the French were the first colonisers, just before 1600. Saint Lucia remained firmly a French possession until about 1750, when the British became interested.

During this period of over 150 years the distinctive culture of Saint Lucia was formed, and the patois language developed. Administration was based in Soufriere, centre of the plantation economy and a port. Castries was

ISLE DE S^{TE}. ALOUSIE

unimportant at this time, the other major ports being Dauphin, Praslin and Micoud.

In 1745 the French began to divide the island into *quartiers,* and by 1780, these had developed into eleven parishes. The *quartiers* became the basis of local administration and were retained by the British.

29

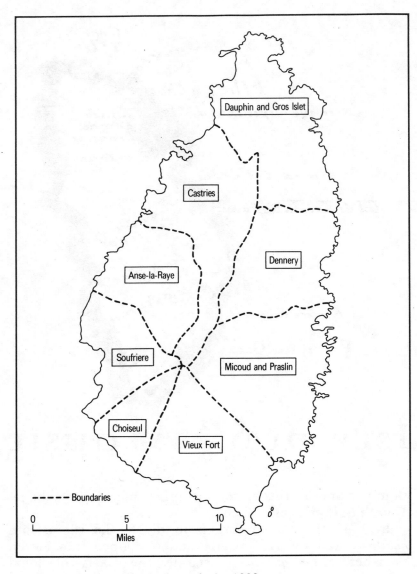

Dauphin and Gros Islet

Castries

Anse-la-Raye

Dennery

Soufriere

Micoud and Praslin

Choiseul

Vieux Fort

- - - Boundaries

0   5   10

Miles

Fig. 19 St Lucia: quartier boundaries 1888

The French Revolution and the Napoleonic Wars   The French Revolution began in 1789. The monarchy was overthrown and a republic set up. Very soon, this republic, under the leadership of Napoleon Bonaparte, tried to increase the power of France, and a great series of wars resulted, lasting until 1815. Between 1760 and 1830 Saint Lucia changed hands fourteen times, and the Napoleonic Wars had several lasting effects on the island

1   The sugar trade was interrupted, and declined. Whereas in 1780 there were over 100 mills on the island, in 1820 only 35 remained in production.

2   Napoleon introduced a new system of land inheritance to all French areas. It was called the Code Napoleon, and has been a major hindrance to agricultural development in modern times.

3   The capital city was moved from Soufriere to Vigie, then to Castries just before the wars really began. This was to make use of the strategic value of Morne Fortune and Port Castries, since invasions came usually from the north.

British Saint Lucia

Saint Lucia became the major Caribbean prize in the sea battles of the Napoleonic Wars. It was from this island that the British fleet sailed under Admiral Rodney to defeat the French at the Battle of the Saints in 1782.

The French governed all their colonies by direct rule, which meant that French culture was imposed. By contrast, the British used indirect rule, keeping control, but allowing local culture and institutions to develop. This explains why so much of the cultural and legal character of Saint Lucia remains French to this day.

Slavery was abolished in 1838, but it took some time to

adjust to the new type of society. The Negro Saint Lucians still had little choice of livelihood. There was poor peasant farming on the hills or poor plantation work in the valleys. Saint Lucia became a Crown Colony within the British Empire. Under this system each colony or territory developed a different political system.

Gradually a minority of Saint Lucians were beginning to share in the running of the colony. The local political patterns that emerged were different from those of nearby Grenada or Saint Vincent. The national character of Saint Lucia was emerging in a formal way and in 1924 a new Constitution and Legislative Council system were established. Internal self-government began to appear.

The British Empire spanned the world and was based on control of the oceans. Castries became one of the most important coaling stations for the British navy. The barracks on Morne Fortune, begun by the French, were greatly extended for fear of new French invasions from Martinique at the turn of the century.

With the death of Queen Victoria in 1901, relationships between Britain and France improved and the final designs for the Morne Fortune fortifications were never carried out. Temporary revivals of Saint Lucia's strategic importance occured during both World Wars, (1914-18) and (1939-45). During the second, military airports were built by the U.S.A. near Gros Islet and Vieux Fort.

Independence

After twenty years of discussions, the British Government decided to give independence to her West Indian colonies in the form of a Federal Nation. This was in 1958. Saint Lucia was well prepared, having begun a ministerial system of government in 1956. Unfortunately the scheme did not work, and the Federation split up, Saint Lucia becoming, in effect, a colony again in 1962.

Associated Statehood

In 1967, Saint Lucia became an Independent State in Association with the United Kingdom. This means that Britain looks after foreign affairs and defence, but otherwise Saint Lucia is fully independent.

Associated Statehood required a new Constitution, under which Saint Lucia gained her first Premier, John Compton, and first Governor, Sir Frederick Clarke. The new Constitution of 1967 included many important political changes

1 A House of Assembly was formed. This is the Parliament of Saint Lucia and consists of one elected member for each constituency, or voting area. In addition there are a small number of nominated members.

2 In place of the old Executive Council, a Cabinet was formed. This is the government of Saint Lucia. Members of the Cabinet are chosen by the Premier, and are usually given a particular Ministry to look after.

3 The Opposition. In any independent democracy there is always some official and open opposition to the Government. In order that this genuine and proper opposition should have a voice in the House of

Assembly, the Governor appoints an official Leader of the Opposition.

4　A Governor now replaced the former official who had been known as the Administrator. This Administrator had always been a British civil servant. Since 1967 there have been three Governors all, of course, Saint Lucians. As Saint Lucia is still linked to the United Kingdom, and is a member of the British Commonwealth, the Governor is the representative of the Queen. He is chosen on the advice of the Premier of Saint Lucia.

Full independence

Late in 1975, it was announced that Saint Lucia would proceed to full Independence in 1976.

The Law

In every society or nation there is often a clash between the need for individual freedom, and the need for co-operation.

It is necessary, therefore, for the Constitution of any nation to make provision for a proper legal system, that will help to reconcile these needs. In the words of the Constitution of Saint Lucia, . . .'every person in Saint Lucia is entitled to the fundamental rights and freedoms. It goes on to list these fundamental rights and freedoms as

1 Life, liberty, security of the person and the protection of the law.
2 Freedom of conscience, of expression, of assembly and association.
3 Protection for the privacy of his home and other property and from deprivation of property without compensation.

Thousands of detailed rules and regulations have been made over the centuries, and they go together to form the 'law of the land' in Saint Lucia. In practice the law consists of the legal institutions of the State, such as the courts, and the independent practitioners of the law, the lawyers.

There are two very important institutions of the law of Saint Lucia that are, in fact, people

1 The Attorney General is in charge of prosecutions. His decisions are independent of the Governor and of the Premier and House of Assembly, because these people must not be 'above the law'.
2 The Judge is responsible for the conduct of the Courts, where prosecutions are heard, defence is heard, and decisions of punishment are taken.

Saint Lucia's Courts are part of the West Indies Associated States Supreme Court, and every one on the island, whatever their rank or fame, is subject to the law.

The Public Service

A nation holds itself together by the services some members of the community provide for others. We call these people public, or civil servants, but they do, of course, get paid for their work.

Civil servants are employed in, or by, one of the Ministries of the Government. They range from highly qualified and experienced experts to manual workers but they all have a part to play. Let us take three areas of concern for the public service.

The health of the nation In modern societies it is thought to be a good thing to make medical care available to all. Such services are provided by free public services through the health centres and hospitals and by fee-paying private services through the general practitioners. Obviously those doctors in private practice are not civil servants, but they do work within the national regulations for medical provision in Saint Lucia.

Also essential are the pharmacists who prepare the medicines. Like the private doctors they are a commercial service, but vital to the maintenance of the health of the nation. Behind the scientific skills of the experts there must be the backing of good nursing. It is the nurses who take the strain of the day to day work in hospitals and health centres. Saint Lucia's nurses are trained at Victoria Hospital in Castries, under the supervision of the matron.

The knowledge of the nation All over the world there has been an 'explosion' of knowledge in recent decades. This means that what is taught in schools can become out of date very quickly. It is not surprising to find that, in the 1970's, the education system of Saint Lucia is changing.

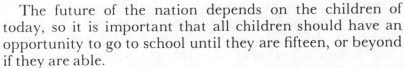

The future of the nation depends on the children of today, so it is important that all children should have an opportunity to go to school until they are fifteen, or beyond if they are able.

Primary education has long been established in the island, thanks mainly to the early work of the Church. Now the Government is trying to provide wider opportunities in secondary and further education. For example, Junior Secondary Schools have been established to increase the quantity of education available to the 12-15 age group.

The Morne Education Complex makes intelligent use of the old barracks to house several colleges. There is a teachers' college, shown above, technical teachers' college, technical college, catering school, university centre (U.W.I.) as well as international foundations like the U.N. Planning School and the Rockefeller Bilharzia Research Centre.

The arteries of the nation Development means movement. Patients must get to hospital, students to college, bananas to the boat and vegetables to the market. Even ideas and instructions must move; by private delivery, public post, telegram or telephone, radio or television. This whole area of activity may be called 'Transport and Communications', about which more will be said later.

The Ministry of Communications and Works in Saint Lucia has special responsibility for maintaining the roads of the nation and the public buildings. Roads may be likened to the arteries of the human body. If they are blocked or damaged, the pulse-rate of the nation will slow down. The public buildings are like the organs of the body, the centres from which Government carries on its work.

Citizenship

The Government operates the Constitution and organises the public services on behalf of the people of Saint Lucia. It does this with money provided by the people in the form of taxation. For example in 1971 the Government expenditure was about twenty million dollars EC.

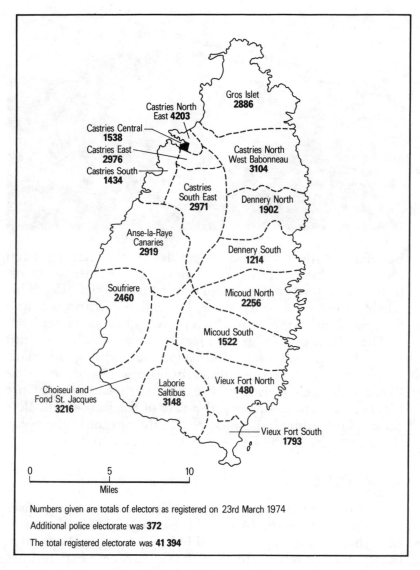

Fig. 20 Constituencies for the 1974 General Election

Saint Lucia is a democratic nation, which means that free elections take place under the rules of the Constitution. Such elections give opportunities for any adult citizen to seek public office. Citizenship is not only concerned with the safety of its members. It also includes the concept of participation by the citizen.

Habits of good citizenship do not have to wait until voting-age is reached. They can be, and are, learnt in the family, the school and the neighbourhood. Indeed anywhere where some sort of community exists. A healthy national community is likely to be one where local councils and other smaller organisations are active and strong. This is citizenship at the local level.

As the population grows and the pattern of settlement changes, it is vital to ensure that the people are being properly represented in the House of Assembly. From time to time a review is made of the constituency areas, and for the General Election of 1974 the number of constituencies was increased, to seventeen. Such an increase should make for stronger local involvement in national administration.

Key words

independence	amerindian	ministry
court	nation	culture
government	rights	state
parishes	elected	judge
administration	monarchy	nominated
lawyer	tribal	republic
council	health	colonial
strategic	constituency	education
federation	constitution	cabinet
transport	opposition	communications

Economy 1

The main task of the government of any country is to look after the economy. This means taking an interest in production, consumption and exchange of goods. In some countries the government controls the economy completely, while in others it merely helps to regulate it. The study of economy is called **Economics.**

If the economy is developing well then more money can be taken out, by taxation, to provide Social Services for the citizens. However, the first objective of the economy is survival. This means obtaining sufficient food. This can be done by growing or rearing it, by buying it from other countries, or both.

Like many young nations Saint Lucia has an economy based on agriculture. First there is domestic agriculture,

Local agriculture

Saint Lucia is a nation built on agriculture. Most Saint Lucians have some land and know how to use it. Here is a study of a small family farm in Garrand, near Babonneau, belonging to Mrs Augustin.

The Augustin Farm This photograph shows Mrs Augustin with five of her ten children: It is well known that she is one of the best farmers on the island. With her husband, she provides nearly all the family's food from two areas of land: (a) one acre around the house and (b) seven acres on a nearby hillside.

Like many Saint Lucians she inherited the land, and many other members of her family live nearby. The legal occupation of land is called land-tenure. Mrs Augustin has had her land properly surveyed and mapped. This is always a wise thing to do.

Every inch of that farm is most carefully tended. All members of the family who are still living at home help in some way, after school, after work, at weekends and holidays. With all this hard work, three crops per year are obtained from each of the following plants.

Augustin Farm   3 crops per year

Black Peppers	Ginger	Spring Beans
Cabbages	Leeks	Sweet Peppers
Celery	Lettuces	Sweet Potatoes
Chives	Ochroe	Thyme
Christophenes	Paw Paw	Tomatoes
Cucumbers	Pigeon Peas	Water Melons
Egg Plant	Spinach	

many people feed themselves from their own land. Then there is export agriculture; bananas and copra.

Next comes the development of manufacturing which needs reliable supplies of power and water.

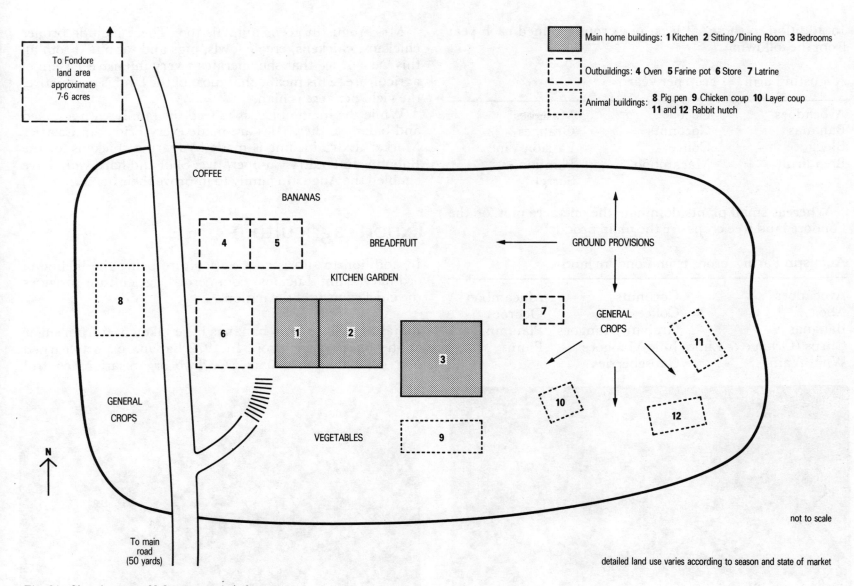

Main home buildings: **1** Kitchen **2** Sitting/Dining Room **3** Bedrooms

Outbuildings: **4** Oven **5** Farine pot **6** Store **7** Latrine

Animal buildings: **8** Pig pen **9** Chicken coup **10** Layer coup
11 and **12** Rabbit hutch

To Fondore
land area
approximate
7·6 acres

COFFEE

BANANAS

BREADFRUIT

GROUND PROVISIONS

KITCHEN GARDEN

GENERAL CROPS

GENERAL
CROPS

VEGETABLES

N

To main
road
(50 yards)

not to scale

detailed land use varies according to season and state of market

Fig. 21   Sketch-map of Mrs Augustin's farm

In addition to these, at least one crop is gained each year from the following.

Augustin Farm   1 crop per year

Avocadoes	Cocoa	Mangoes
Bananas	Coconuts	Oranges
Bloggose	Coffee	Passion Vine
Breadfruit	Macamboo	Plantain
		Sorrel

Whereas small plants dominate the one-acre plot, on the Fondore land tree crops are the main product.

Augustin Farm   crops from Fondore land

Avocadoes	Coconuts	Macamboo
Abe	Coffee	Ochroe
Bananas	Graham Mangoes	Plantain
Citrus (Orange & Lime)	Julie Mangoes	Plums
White Yam	Gooseberries	

Mrs Augustin keeps animals too. These include broiler chickens, chickens, creole fowls, pigs and rabbits. From all this we can see that she operates a very intensive system of agriculture. This means that most of the land is productive, the yield per acre is high.

While the main object is to feed the family with a mixed and balanced diet, sales are made every Friday in Castries Market. Extra income is made by rearing chickens for the Babonneau Poultry Co-operative. Skill and hard work have enabled the Augustin family to improve their home.

Export agriculture

In addition to subsistence agriculture being the lifeline of ordinary Saint Lucians, commercial agriculture provides the chief exports of the nation.

Bananas   Since 1956 bananas have taken over from sugar as the main export crop. In 1972 banana exports earned over eight million dollars EC. Two organisations control

the banana trade of Saint Lucia
1   Saint Lucia Banana Growers Association.
2   Geest Industries.
All bananas for export must be sold to the S.L.B.G.A., who then sell to Geest for transportation to Britain.

Banana plantations are of all shapes and sizes. This one is in the Roseau Valley. A similar plantation covers most of the Cul-de-Sac estate near Castries. This estate is divided

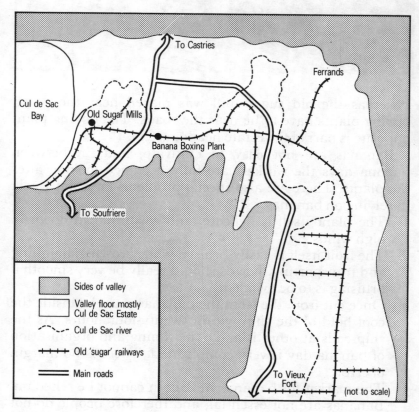

Fig. 22   Cul-de-Sac Estate

into two working areas, each of which has about forty-five workers and a manager. Each banana plant produces only one crop, and so, in order to maintain production, the estate has plants in various stages of growth.

'Banana Day' occurs once per week, when the boat docks at Vieux Fort and then Castries. Mature bananas are selected, and the bunches are cut into hands on the spot. The hands are placed on a tray which is headed across the drainage ditches to the railway wagon. These little railways, built for sugar cane export, are then used to transport heavy loads of bananas to the boxing plant.

The boxing plant is the focus of the estate. Here the bananas are washed, checked, sprayed with a chemical to dry up the sap, and packed into cardboard boxes. All this takes place in a specially designed building.

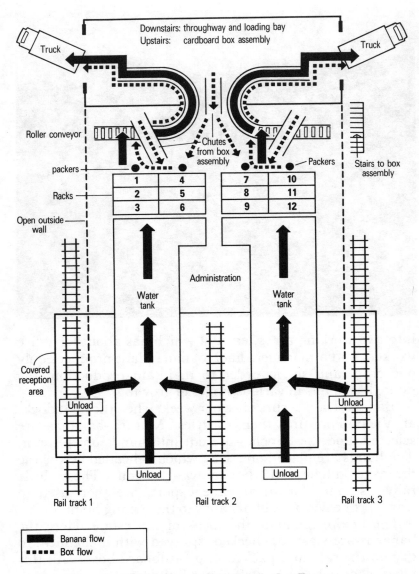

Downstairs: throughway and loading bay
Upstairs: cardboard box assembly

Truck

Truck

Roller conveyor

Chutes from box assembly

packers

Packers

Stairs to box assembly

Racks

1	4
2	5
3	6

7	10
8	11
9	12

Open outside wall

Water tank

Administration

Water tank

Covered reception area

Unload

Unload

Unload

Unload

Unload

Rail track 1

Rail track 2

Rail track 3

| Banana flow |
| Box flow |

Fig. 23 Plan of main boxing plant at Cul-de-Sac Estate

Whereas the old sugar mill was placed near the coast, boxing plants have to be near the roads. The boxing plant opposite is part of the estate headquarters.

Bananas are not easy to cultivate, and, in between banana days the workers are fully occupied looking after the plants. There are also special problems for Saint Lucia in relying on bananas.

1   The plant itself is delicate, and easily blown down in high winds.

2   The fruit itself is easily bruised while growing, handling and transporting. Roads should really be very smooth if bruising is to be avoided.

3   Once cut from the stem the bananas must be inside the cool hold of the ship within twenty-four hours. As the ship calls at other islands, the timing and organisation of banana day is very complicated. There is no margin of error.

4   The market for bananas in Britain cannot be relied on. Bananas are not essential, and therefore people do not need to buy them. Bananas are expensive in Britain.

Transport costs are high. But despite these problems, Saint Lucia must depend on bananas for years to come.

As this map shows, the growing, boxing and shipping of bananas is now a major geographic feature of Saint Lucia.

Copra   This is the second export of Saint Lucia, bringing in 270 000 dollars EC in 1972, and is easier than bananas in several respects.

1   Coconuts do not only produce copra, there is also coconut oil, worth nearly two million dollars EC in 1972, and coconut meal, worth 150 000 dollars EC in 1972.

2   Unlike bananas, there is little wastage. Coconuts are more useful. At the processing plant near Soufriere, the

1 & 2   Management Houses

3   Copra oven

4   Copra bagging and storage

5   Vehicle ramp

6   Piggery

7   Stables

8   Chemicals

9   Boxing plant process building

10   Workshop

11 & 12   Fertiliser stores

13 & 14   Fuel tanks

++++   Railways

═══   Roads

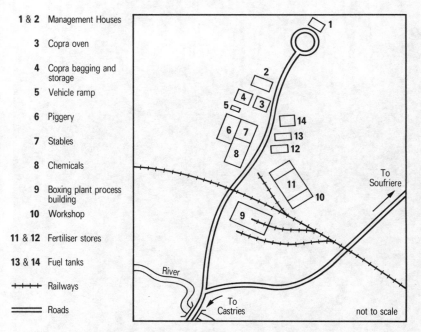

Fig. 24 Boxing plant complex at Cul-de-Sac Estate

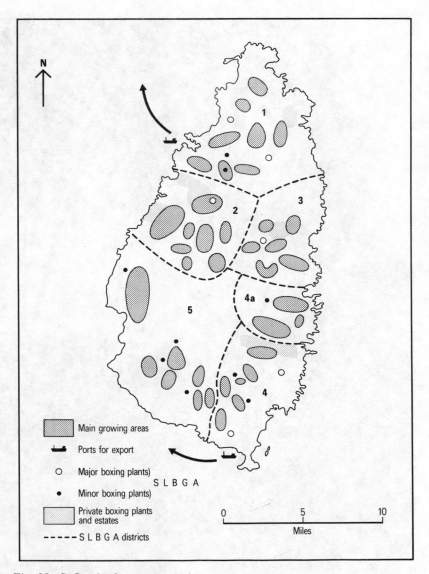

Main growing areas

Ports for export

○ Major boxing plants)

● Minor boxing plants)   S L B G A

Private boxing plants and estates

– – –   S L B G A districts

0   5   10
Miles

Fig. 25 St Lucia: banana growing areas

43

various bi-products are manufactured, for example soap.

3   Copra is exported to other West Indian countries in the CARIFTA group, and is not, therefore, dependent on such a distant market as bananas.

Like bananas, copra is marketed through a national co-operative, which is called the Saint Lucia Copra Growers Association. This particularly helps small farmers to gain some income from coconuts.

Other crops   In the early days of Saint Lucian commercial agriculture, there was a much greater variety of products than today. At present, in addition to bananas and copra, only very tiny quantities of cocoa, fresh fruit and vegetables are exported.

Problems for Saint Lucian agriculture

Agriculture is the 'life-blood' of Saint Lucia, but it faces many very difficult problems.

1   Commercial agriculture, which provides over 90% of the nation's visible exports, depends too much on one crop. There is a great need to diversify crop production. In particular, great numbers of citrus fruit trees have been neglected. Hotels may well provide a larger market for local produce.

2   The system of land tenure inherited from the French colonial period is leading towards a fragmentation of holdings. Farms become very small and commercial production inefficient. Such a system of land inheritance discourages incentive, and much land has gone out of production because of this.

3   Young people are leaving rural areas. This is called rural depopulation. It often means that not enough heavy work is possible because only the old and very young remain. Also, traditional skills of local farming are not being passed on effectively.

Water and power

Water   Without water neither people nor industries can survive. Saint Lucia has plenty of water through her rainfall, but supplies to buildings are often limited. Not enough is done at present to capture the natural supply before it runs off into the sea. Being a small, hot island, Saint Lucia can lose its moisture very quickly.

If reservoirs were to be built in the main water catchment areas, a more reliable water supply could be provided. Every year there is a bigger demand for water as the population increases and more houses are built. New hotels and industries must have reliable water supplies to succeed.

Sometimes it is necessary to provide extra water for the soil. This is called irrigation, and could be very useful in the drier parts of the island.

Many diseases can be carried in water. If clean water could be piped to each home then people would be able to keep out of the rivers, and so avoid bilharzia.

Power   Power is to industry and transport as blood is to the human body. There are sources and types of power.
1   Sources of power are natural, for example coal.
2   Types of power are man-made like electricity.
We should never forget, of course, the most important power supply of all, the muscles of the human body, and of our animals, but this is an unusual use of the word 'power'.

In Saint Lucia, petroleum is the main source of power. This is usually drilled-for and refined in Trinidad, by several oil companies. Tankers bring different types of

petrol and oil to the island. Storage depots are located at Castries, Soufriere and Vieux Fort. These supplies serve all the road transport on the island, and many of the boats too.

Gas is found with, or made from, petroleum. It comes by tanker from Trinidad and Antigua, and is very important for domestic cooking purposes.

Geo-thermal sources of power exist in Saint Lucia as is shown in the photograph opposite. This means underground heat. By sinking a shaft deep into the earth near the Sulphur Springs, superheated steam could be obtained and converted into electricity. If fresh supplies of water are injected, the process can be continued as long as volcanic activity exists. There is a plan under way to do this, which would give Saint Lucia an abundance of power.

Hydro-electric power, which means electricity made from the power of running water, is already produced on a small scale near Soufriere. Many areas of Saint Lucia have abundant continuous running water, so hydro-electric power supplies could be increased. Once installed, and in operation, this is a very cheap source of electricity, and is available as long as the rain falls and the rivers flow.

All the electricity produced is fed into the electricity grid. This is a network of cables carrying power to offices, shops, hotels, industries and houses. People pay for whatever they use, at a rate charged by the Saint Lucia Electricity Board. Most of the power is generated at the Union Power Station.

Industries

Power and water supplies help people to work more easily and with greater effect. The word 'industry' really means people at work, and this includes agricultural work. There are three types of industry.

1   Primary industry which means working to obtain the raw materials of nature, by farming, fishing, forestry, mining and quarrying. The last two are called 'extractive industry'.

2   Secondary industry which means making things from the raw materials of primary industry. It is often called manufacturing industry, for example making furniture or beer.

3   Tertiary industry which means providing a service such as banking, shopkeeping, teaching, running a hotel.

The more economically advanced a nation is, the more tertiary industries it tends to have.

Manufacturing industry   This is the type of industry that makes a country economically strong. Unfortunately small nations like Saint Lucia do not usually have the raw materials to make many things from. Neither do they have a big enough home market in which to sell a sufficient quantity of anything to make a profit.

Nevertheless there are a number of successful small industries on the island which are based on imported materials, such as furniture and ironwork products on the Morne, and soft drinks in Castries.

We may take another industry in Castries as a case-study to illustrate just what is involved in manufacturing.

Ferrands Dairy and Saint Lucia Cold Storage   This concern consists of two companies, one of which is subsidiary to the other.

1   Structure.   Industries are groups of people. The usual way to start an industry is to form a company. This consists of a board of directors, and a number of shareholders, who each provide a share of the money, called capital. They have made an investment. The company must have a labour force, so it employs people.

2 Factory. Manufacturing requires a special building and equipment. The building, called the workshop or factory consists of various spaces and machines. All this is called the plant.

3 Materials. Ferrands Dairy produces reconstituted milk and ice-cream. For this process the company requires a number of materials, which have to be obtained from other countries.

4 Marketing. The products of a company may be sold direct to the public, which is called wholesale, or through other shops, which is called retail. Both methods are used in this case.

Fig. 26 Plan of Ferrands Dairy Plant

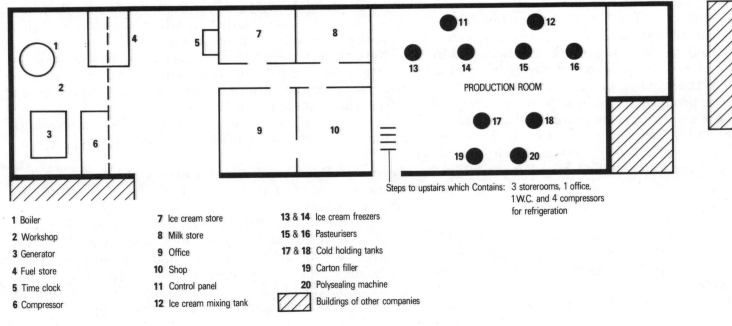

PRODUCTION ROOM

Steps to upstairs which Contains: 3 storerooms, 1 office, 1 W.C. and 4 compressors for refrigeration

1 Boiler	**7** Ice cream store	**13** & **14** Ice cream freezers
2 Workshop	**8** Milk store	**15** & **16** Pasteurisers
3 Generator	**9** Office	**17** & **18** Cold holding tanks
4 Fuel store	**10** Shop	**19** Carton filler
5 Time clock	**11** Control panel	**20** Polysealing machine
6 Compressor	**12** Ice cream mixing tank	▨ Buildings of other companies

A Income

Ferrands Dairy		St Lucia Cold Storage	
50%	Ice Cream	40%	Fish
40%	Milk	40%	Poultry
10%	Small Items (cones)	10%	Meat
		10%	Vegetables

B Expenses

Ferrands Dairy		St Lucia Cold Storage	
50%	Raw materials (Fat, milk powder, plastic bags, cartons, ginger, ice cream cartons – mostly imported	80%	Raw materials (Fish, poultry, meat, vegetables(– mostly imported.
		5%	Salaries and wages†
25%	Salaries and wages.	5%	Electricity
10%	Electricity	5%	Maintenance
10%	Maintenance	4%	Depreciation*
4%	Depreciation*	1%	Water
1%	Water		

*Depreciation means a loss in the value of assets, e.g. machines.
†A low figure due to some overlap of staff with the other Company.

5 Finance.   The object of the exercise is to get back more money from sales than you spend on materials, wages and other expenses like water and electricity. If you do this you make a profit, which may be re-invested, to improve the company in the future. Here is the way in which our sample companies gain and spend their money.

Key words

land-tenure	petroleum	irrigation
services	intensive	refining
drainage	raw materials	yield
storage	reservoir	home market
income	gas	primary industry
export market	subsistence	geo-thermal
secondary industry	factory	plantation
hydro-electric	tertiary industry	company
boxing plant	electricity	manufacturing
shareholders	market	grid
machinery	capital	export
power station	profit	investment

Economy 2

While power and water supplies are essential for production, transportation is essential for trade. In modern and developing societies there tends to be specialisation. Some places have industries, others have not. Goods are therefore moved from place to place where they are bought and sold. This process is called trade.

In other words, whereas 'primitive' communities may be self-sufficient, 'advanced' societies tend to be interdependent. This means that they are linked together and depend on each other.

An obvious example of this is the development of commercial services. Many such services are concerned with the handling of money, the medium through which modern economies are regulated. Think of the many banks,

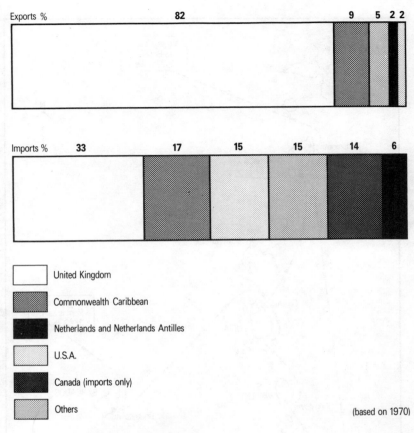

Exports %

| 82 | 9 | 5 | 2 | 2 |

Imports %

| 33 | 17 | 15 | 15 | 14 | 6 |

□ United Kingdom

▨ Commonwealth Caribbean

■ Netherlands and Netherlands Antilles

▨ U.S.A.

■ Canada (imports only)

▨ Others

(based on 1970)

Fig. 27 St Lucia: external trade

insurance companies and trading companies in Castries. Garages, hairdressers, and buses are examples of another type of service where money is exchanged 'for services rendered'. Shops are the modern equivalent of markets.

Some services can even be exports. Tourism is now a major component of the Saint Lucian economy. Provided that such services can be maintained they will help to diversify the economic structure of the island.

External trade means trading with the outside world. For Saint Lucia, being an island nation, this depends almost entirely on ocean shipping, with some air freight too.

Goods coming in are called imports, while those going out are exports. A study of the origins of imports and the destination of exports gives the geographic pattern of trade. Usually in any country there is a difference between the value of its imports and exports in any one year. This difference is called the **balance of trade.**

But in comparing exports and imports we are dealing only with goods, and so we call any difference the balance of visible trade. For Saint Lucia, the imports are worth more money than the exports, so there is a loss on visible trade. However, this is partly made up by money gained from tourism which is a type of invisible trade. Customs duty is another example of an invisible trade item.

Transport

Ocean shipping The inter-island schooners and deep sea cargo boats are traditional features of the Caribbean scene. Before the motor age there was a lot of travelling by coastal boats between the seaside villages of Saint Lucia.

Schooners may be fewer than in the past, but cargo vessels from Europe and North America are still the lifeline of the island. Saint Lucia is a developing country, but, being small, she has still to import many things. Some obvious examples are flour, petroleum, and many household goods from freezers to saucepans.

All items of machinery, like air conditioners, typewriters and motor vehicles, have to come by boat. The Geest boat, shown overleaf, is unloading bananas at Barry in South Wales. When it returns to Saint Lucia it will bring many

manufactured imports from Britain. Every week one of these boats calls at Castries to unload goods, and a few days later to load bananas.

Land transport Without some method of transportation most human activities would have to stop. The most basic form of transport is to use human and animal muscle power, i.e. to walk. People do not usually move all over an area of land. They keep to the best routes. These become worn away as tracks. At a later date some of these tracks may be properly surfaced to become metalled roads.

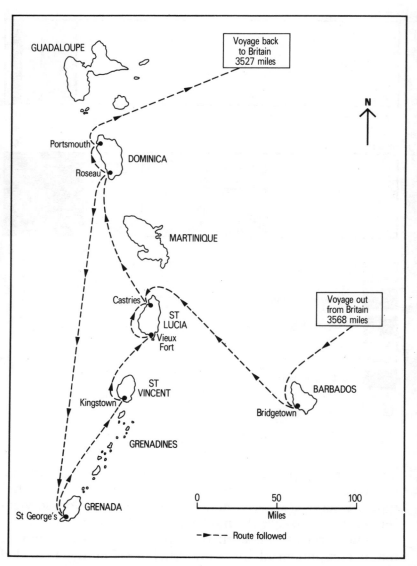

Fig. 28 Geest boat itinerary

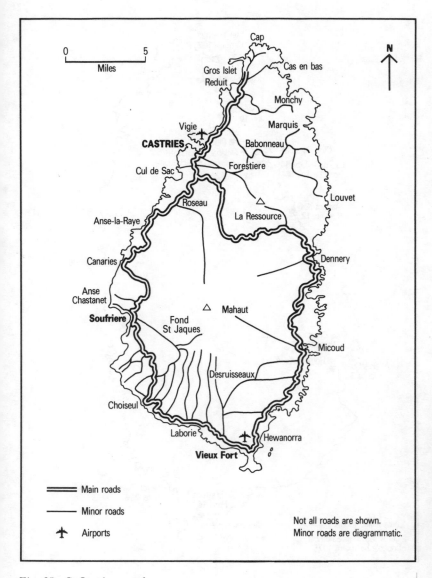

Fig. 29 St Lucia: roads

The old plantation system of the island was based on mule transport. Mules could be used in the steep interior. Motor transport cannot reach such places, and so, citrus fruit production in particular has declined. In some of the large flat valleys narrow-gauge railways still exist. These were built for the export of sugar, and are unimportant today, though still used on some plantations for carrying bananas to the boxing plant. Nowadays the basic land transport medium is the road system.

The buses of Saint Lucia have replaced the old system of boats between villages. Most buses make one return journey per day, to Castries and back. In recent years minibuses have been constructed from small open vans. These make more frequent and shorter journeys than the larger buses and are much in demand. In time it is likely that a regular system of bus routes will develop. Already some properly constructed minibuses are being used.

Road transport is one of the essential factors in developing a modern society. Like good electricity and water supply the roads are part of the infrastructure of the community. They have to be surfaced and maintained.

Lorries are vital to the transporting of bananas from the boxing plants to the ports. Bananas bruise easily on bumpy journeys and thereby lose value.

Air transport Saint Lucia has two airports, Vigie near Castries and Hewanorra near Vieux Fort. Four types of air traffic operate:

1 Major world airlines, like British Airways and Eastern Airlines, call at Hewanorra.
2 Caribbean airlines, like Liat, operate through Vigie.
3 Local companies like Wings operate to other islands and within Saint Lucia.
4 Private aircraft are based, or call, here.

Air transport has proved particularly useful in making Caribbean commercial and political meetings much easier to organise than hitherto. It is also very important for the tourist trade.

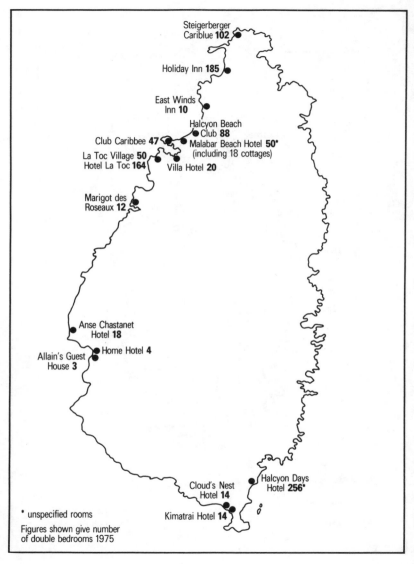

Fig. 30 St Lucia: hotels 1975

Tourism and services

One of the advantages of a society becoming more developed is that a wider variety of jobs is created. More shop assistants are needed, bank clerks, typists, and many others. All this means a higher income in the family and possibly a better standard of living. These people are providing services, and making money at the same time.

Tourism and Saint Lucia We have already seen that when a tourist spends money in Saint Lucia, we call it an invisible export. This means that money has come into the island but nothing has gone out. In fact the island is 'exporting' its sun, sea and sand.

In recent years, tourism has earned Saint Lucia about the same amount of money as have banana exports. It is changing the appearance of many parts of the island. The most obvious example of this is the construction of large hotels. In 1974, the pattern of hotels and guest-houses was as shown opposite.

Clearly the Castries area is the main focus. This is partly due to the fact that the hotels need constant supplies of goods and services which are more easily available in the capital city. Also, many tourists like to buy souvenirs from the shops in Castries. The development of these hotels has provided employment in many ways. First the hotels had to be built, together with various services leading to them such as roads, water, electricity.

Then the hotels have to be staffed. They need cooks, waiters and waitresses, receptionists, cleaners, secretaries. Such jobs were not available to Saint Lucians before about 1965. Now a catering and hotel school has been opened on the Morne Educational Complex. School pupils are learning about tourism in their Social Studies classes.

HOTEL

SUPPLIES What does it need?	TRANSPORT How do people get there?	RECREATION What do the visitors do?	THE SITE Where is it and why?	THE BUILDING What does it look like?	ADMINISTRATION How does it work?
Basic	Local	Sports	Geography	Interior and	Advertising
Water	*Taxis etc.*	*Fishing*	*Physical site*	Exterior	Management
Electricity	External	Music	*Land values*	*Architecture*	*Structure of*
Human	*Airlines, ships*	*Disco*	History	*Design*	*departments*
Labour	*and tours*	Bar	Economics	*Efficiency*	*Organisation*
Food		*Rum*			*Personnel*
Materials (hard		Visits			
and soft)		*Sulphur Spring*			

A hotel is a complicated organisation and can make an interesting case-study of a service industry.

There are even opportunities for entertainers; singers, dancers and musicians. People often do these as part-time jobs at weekends and in the evenings.

Different types of tourism exist in Saint Lucia as in other Caribbean islands.

1 Stopover tourists. These are the visitors who stay in hotels. The length of stay varies from one night to several months, but the average is about five to seven days. Of course, some of these people are not really tourists; they may be business men or visiting officials from other countries, but they are recorded as tourists because they have used hotel accommodation. In recent years, about 50 000 people per annum have stayed in Saint Lucian hotels. In 1964, for example, there were only 8500. In that year Grenada had over 11 000 and

Saint Vincent about 12 000, whereas their numbers in recent years have risen only to about 30 000.

In other words Saint Lucia has built more hotels, but for most of the year the occupancy rate is low. For example, in 1972 only one-third of the rooms available were occupied. It is likely that hotel building will now proceed at a slower rate.

2 Cruise ships. Saint Lucia is a port of call for many Caribbean cruises. Sometimes a cruise may be combined with a week or two at a hotel on the island, but most passengers disembark for one day only. They spend money in the shops and in visiting places, such as Morne Fortune and the Sulphur Springs.

3 Yachts. There are always a number of private and chartered yachts calling at Saint Lucia. They have to pay fees to moor at a harbour and the passengers spend money on the island,

Tourist arrivals from	1973	1974	% change 73-4
U.K.	8 092	13 563	+68%
U.S.A.	11 359	12 039	+ 6%
Canada	8 785	9 844	+12%
South America	1 441	1 311	− 9%
West Indies	12 384	11 142	− 6%
Others	3 748	3 911	+ 4%
Means of travel			
Air	43 902	48 585	+11%
Sea	1 907	3 231	+69%
Total of staying visitors	45 809	51 816	+13%
Total of in-transit visitors	46 485	43 145	− 7%
Total visitors	92 294	94 961	+ 3%

Money spent by visitors in 1973. 4 355 997 $ EC
Source: *West Indies Chronicle* Saint Lucia Supplement May/June 1975.

We shall consider tourism again in the final chapter of this book which is about 'development'.

Commercial and financial services Tourism is the most recent and dramatic example of a commercial service in Saint Lucia. But there are many others that we tend to assume will be there when required. Tourism benefits Saint Lucia economically and is enjoyed by foreign holidaymakers. The services we shall now look at are part of the fabric of modern life on the island and an important aspect of development.

All our activities are related by money. Money is a device or method of placing a relative value on goods and services. Over hundreds of years the idea of money has become so complicated that a whole range of services exist to deal with it. The most obvious of these are the banks.

Traditionally, banks are places to save your money in. It is safer than putting it under your bed. You can open an account by depositing some money, and then you receive a cheque book. Then you can withdraw money from time to time by using a cheque. Cheques also allow you to buy goods without handling cash, but not every shop will agree to receive cheques.

If money is not used, its value tends to decline. This is due to a general tendency towards increased costs, which is called inflation. Banks therefore invest the money they have been entrusted with, and make a profit. This extra money is used to pay their running costs, and improve their services. Banks also lend money to people and charge 'interest' on the loan. This means that the borrower pays back more money than he received, but at least he gets the money when he needs it. Most industries are started with a loan from a bank. Such a loan is called credit.

Another type of financial service is the insurance company. By paying small regular amounts called premiums, you can get back an agreed amount of money in the event of an accident or disaster. All motor cars must be insured.

A co-operative is a sort of bank where local people join their resources together to cut down their individual

expenses. Then, of course, any profits are beneficial to all those who are members of the co-operative.

Somebody has to organise the import and export of goods. Most things sold in Saint Lucia originate from Europe, North America or Japan. Companies in those places appoint local agents to sell their goods on the island. For example all the motor companies have agents, so do breweries, travel firms and insurance companies.

Trading companies often keep large stocks of goods in storage, and may sell direct to the public. This is called wholesale, but they also sell to shops who then retail the goods to customers at a higher cost.

Shops are the most common type of commercial service. We find many different types of shops in Castries.
1   Large department stores such as J. Q. Charles, and Minvielle and Chastanet. These are usually part of a large trading company.
2   Specialist shops such as clothing shops, bookshops, souvenir shops and bakeries.
3   Supermarkets which are modern self-service shops. They may be part of a department store, or a separate organisation for example the Supercentre at Vide-Bouteille.
4   In the suburbs and villages it is more usual to find general stores selling a wide range of goods, even though they are very small.

The list of commercial services in a modern society is endless. Think of travel agents, hairdressers, garages, restaurants and bars, rum shops, photographers, private schools, chemists, cinemas and many many others. With road transport now so important garages have become a vital service feature.

Key words

internal trade	balance of trade	interest
banks	external trade	visible trade
co-operative	accounts	imports
invisible trade	cheque	inflation
exports	infrastructure	credit
premium	pattern of trade	cash
agency	wholesale	retail

Culture

The word 'culture' has several meanings, both social and scientific. The meaning taken here is that of 'local expression of way of life'. Therefore, in this chapter we look very briefly at the ways in which Saint Lucians express themselves that are distinctive to the island. The patois language, the festivals and the religious activities are examples of this. We look also at Saint Lucian writers and artists who are making great contributions to the cultural life of the island through their work.

A virile folk culture develops with the rapid changes of the modern world. In Saint Lucia this is clearly seen in drama and folk music. However, there is a massive invasion of this culture by journals, books, records, radio and television from overseas. A part of this mass media revolution is Saint Lucian, for example local newspapers like *The Voice* and *The Crusader,* but much is foreign. It could kill local culture.

The main difference between the mass media and the folk culture is that in the former, most people are passive but in the latter they are active and participating.

Language

Words are the voice of a culture. It is very difficult for anybody to understand another language as well as his own. In fact languages may divide people much more than differences in wealth and race.

Saint Lucia has two languages. Patois which is the real language of most of the people and English which is used officially. Patois is based on French, and includes many sounds and meanings from different African languages.

At different times in the past there have been a number of misguided attempts to discourage patois speaking, but it is an essential medium for every child's absorption of Saint Lucian culture. At the same time, English is a necessity in school and work. In many countries of the world two or more languages are spoken by most people. In Saint Lucia, the two languages serve different purposes and there is room for them both.

Folk culture

The term 'folk culture' includes such items as songs and dances, ceremonies and festivals, poetry, stories, novels and plays, painting and pottery, carving, sculpture and games. We may call these creative activities; some are individual and private, others are done by large numbers of people together. All are expressive.

Festivals: La Rose and La Marguerite These two festivals are unique to Saint Lucia, both involving many different folk culture activities. The basis of each is very similar and a Saint Lucian would support one or the other, but not both. Indeed there is a strong rivalry between them. They take place on the days of their respective patron saint:

1 St Rose of Lima 30 August
2 St Marguerite 17 October

The festivals take the form of an elaborate dramatic performance by the local community. Different people take the parts of a royal family, for example, king, queen, prince, princess. The figures seem to derive from the type of monarchy common in eighteenth century Europe, and the festivals relate clearly to the French colonial period on the island.

Other local people take the parts of officials such as the magistrate, chief of police, commander-in-chief and doctor. There is a chantrel appointed to lead the singing, which is a very important element in these festivals.

During the months leading up to the festival day the various performers and groups engage in friendly rivalry and banter, which is also to do with raising money. Seances are also held, often on Saturday evenings, to raise money. The outcome of all this build-up explodes on the day in a flamboyant display of costumes, parades and singing. The players tour the district. Much food and drink is consumed. Virtually everybody in the community has been involved in some way, and the production has been local in every way.

Songs   Much music is made in the process of festivals like La Rose and La Marguerite, and on other days too. However, as elsewhere in the world, genuine local music is being rather pushed aside by 'pop'. Of course, much of the popular music here is of Caribbean origin, and therefore represents a regional folk culture.

About 1970 an attempt was made to revive traditional songs of Saint Lucia. In particular a group of singers called the Hewanorra Voices made a record called 'Authentic Songs of Saint Lucia' which includes the following titles:

Cock Chantez	Femme Blanc	Ti Mamai
Manser Marie	Estephan	Ti Piment
Se Mwen Sho	Al Doo Doo	Tindai

The recording of traditional music is becoming popular now in many countries. It is also necessary for the historical record, as only the old people tend to support traditional music, and it is in danger of dying with them. So there are probably many other songs to be discovered and recorded in the out-districts of Saint Lucia.

Writers

Interest in the folk-culture of Saint Lucia was stimulated greatly by the performance in 1972 of Roderick Walcott's play *Banjo Man*. This drama, based on the local festivals, and with music, was the great success of Carifesta, the Caribbean Festival, held in Guyana in 1972.

Banjo-Man showed that folk culture and popular culture need not be separate. Not all of Roderick Walcott's work is locally based, he is an established playwright in international circles. Then there is Roderick's brother, Derek Walcott, possibly the best poet produced by the West Indies, who also has gained great respect for his work abroad. Poems such as 'The Death of a City by Fire' and 'Dennery' are examples of local writing of the highest order.

Derek Walcott is also an established novelist, as is Garth St Omer. All these men, and others less famous, are artists of creative writing, of drama, poetry and fiction. There has been, in recent years, a great increase in the writing of short plays and poetry. Many of these deal with current problems of Saint Lucian society, and speak on behalf of many more people who do not have the gift of expression. We should not think that we can learn more about a society from Social Studies, than from Literature.

Art

Painting has been greatly encouraged in Saint Lucia by Dunstan St Omer. In addition to creating many fine canvasses he has worked through teacher training, in schools, communities and churches to encourage painting by 'ordinary' folk. In view of the linguistic problems mentioned earlier, his popularisation of the painting as 'communication' has great educational significance.

Through church attendance in large numbers, many Saint Lucians are expressing a particular part of their culture, or way of life. The various churches on the island, have contributed immensely to the education and welfare of the people of Saint Lucia, especially during the colonial periods when both the French and British governments did very little in this respect. Whatever spiritual service the churches may provide for the individual, they also act as a general conservative force in society. This means that they provide a measure of stability and continuity in a rapidly changing world. It is very important not to overlook the immense influence which the churches in Saint Lucia have on many different facets of society.

There has also been encouragement of wood carving, sculpture, pottery and weaving. These are not traditionally strong features of local culture in Saint Lucia, but an effort has been made to revive some of the skills of the past. A handicraft centre has been established at La Fargue, where training in these arts is provided. As more and more people take advantage of this opportunity, this centre may well, like the inspiration of St Omer, stimulate different forms of expression, and strengthen the folk culture of Saint Lucia.

Religion

Religion, in various forms, is one of the deepest and most important features of the culture of Saint Lucia. Some aspects of the religious structure of the society have already been given on page 6. Most of the people are Roman Catholic, the church established by the French, and operated here today by an Order based in Dominica. The Catholic Church is a dominant feature of most Saint Lucian villages.

Carnival Many aspects of folk and popular culture come together in the annual festival of Carnival which takes place in February. The idea of Carnival is found in many catholic countries of the Caribbean and Latin America, particularly where there is a strong African element in the racial background of the people. It is a popular expression of joyful and free activity, providing an opportunity for participation by everybody on the island. Different groups and organisations form 'bands' of performers in costume who depict a theme chosen by them. Each band has its musicians and its followers and the climax of the festival is a colourful parade and competition to find the best bands of the year.

But more important than the organised bands, everybody, whether Saint Lucian or visitor is welcomed in jumping-up and milling through the streets of Castries, beginning with 'jouvez', the early morning of the first day of Carnival.

Although now of obvious tourist significance, Carnival still has great meaning to the average Saint Lucian. There is an opportunity for uninhibited expression, and many strands of historical and present-day feeling are woven into each year's Carnival.

Key words

language	patois	festival
literature	La Rose	mass media
dance	La Marguerite	pottery
drama	Carnival	carving
poetry	religion	sculpture
fiction	music	weaving

Development

What is development? Development means change. Often, but not always, it means progress too. Change has always been with us, from the beginning of the world, and throughout the history of mankind. Everyone is continually changing, and so are the nations they make up.

So development is not new, but what is different nowadays is the very rapid rate of change. Things become out of date very quickly. The problem of development for any country is how best to make use of new ideas, new knowledge, new machines and new methods of working for the good of her own people.

Economic development is needed to increase the wealth of the nation and improve living standards. But economic changes have social consequences, and there needs to be some link between increasing material wealth and preserving the national identity of Saint Lucia.

All this needs careful, yet flexible planning, to make the most of inevitable changes in the future.

Problems and resources

It seems that the more modern a country becomes, the greater the economic problems it faces. Saint Lucia is a trading nation depending on her exports. Unfortunately the economy of the island was founded on the plantation system or monoculture, whereby the island specialised in one or two export crops. Under this colonial economy, linked to France and Britain, industries were not developed. Instead, manufactures were purchased from overseas. Saint Lucia became dependent on imports.

Now that Saint Lucia has Statehood, there is a need to diversify the economy so that it is more balanced. The idea is to develop new exports, in addition to bananas and copra, and also to substitute food, and other products made in Saint Lucia, for those at present imported.

Social problems A nation has to provide for its citizens. This means that the Government tries to develop more services in areas like education, health and communications. Provision of these services costs money; it is part of the economy. If the population increases at all, then each year the Government has to pay more money just to maintain services at the same level!

Problems of resources Resources are the basis of development and they are of two types. Firstly limited resources that cannot be renewed once they have been used up, for example petroleum, iron ore. Secondly renewable resources that can be redeveloped, (provided that great care is taken), such as plants and animals. It is sometimes said that a nation's greatest resource is its people.

The most vital resource of Saint Lucia is the soil, which is a limited resource. For development to succeed it is necessary not only to seek new resources, but also to make better use of old ones. There is a need for conservation, both of natural resources for example fruit trees and of traditional human skills such as farming and fishing.

Land use and development

Saint Lucia is a small island. The limitation of space is obvious. Over the centuries a certain pattern of land-utilisation has developed resulting from the historical plantation economy. This in turn helped to determine the location of settlements.

Types of land-use We can see six major types: (a) Agricultural (farmland and forest), (b) Industrial (mining, manufacturing and services), (c) Commercial (shops, banks, markets), (d) Residential (houses), (e) Recreational (beaches, golf courses), (f) Transportation (docks, roads, airfields). In addition there is unused land.

Obviously, for development to take place certain areas of land-use must change. The present pattern itself is only the result of various changes in the past. It is not enough merely to know the present pattern of land-use. We could ask the following questions.

1   How did the present pattern of land-use come about?
2   What changes can we see taking place today? Why are they happening?
3   What legal and planning processes exist affecting the use of land?

4 What future patterns of land-use are likely to develop?

By far the biggest and most important user of land in Saint Lucia is agriculture. Vegetation and soil are Saint Lucia's most vital assets. The diagram below shows how the four questions could be applied to the problem of agricultural development.

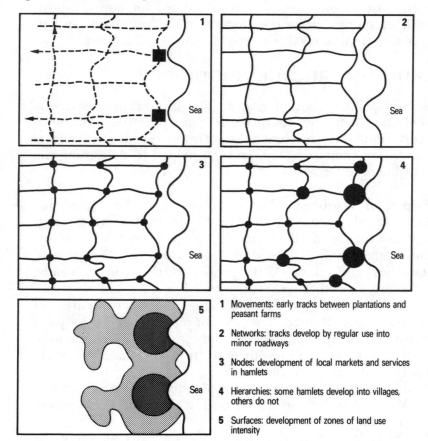

1 Movements: early tracks between plantations and peasant farms

2 Networks: tracks develop by regular use into minor roadways

3 Nodes: development of local markets and services in hamlets

4 Hierarchies: some hamlets develop into villages, others do not

5 Surfaces: development of zones of land use intensity

Fig. 31 A model of settlement development applied to St Lucia

Networks of movement

Different uses of land are made possible by linking them with movements of people and commodities. In other words, some of the land space is used for activities in situ, such as a banana plantation or a house, while other parts of that space are used for activities in transit for example a bus service or a dockside.

Whenever a person moves from one place to another he or she takes a route. It might be simply to walk along a path, or it might be to drive a car along a road, to take an aircraft through the sky, or a boat across the water. These methods form a complicated pattern of networks.

The previous diagram illustrated the development of land use, and included some ideas of networks of movement. Obviously transportation systems have great influence not only on rural land use, but also on urban development. The following diagram shows this influence in general terms and has been applied to the situation in Saint Lucia. Such a model helps us to understand development and to predict what the next stage might be.

The important questions to ask now are:

1 Is there room in Saint Lucia for only one major centre?
2 Is Castries already too big?
3 Are the latest transport developments making Castries more, or less dominating in the island?
4 What will be the future effects of present transport developments?

There have been major developments, extending the networks of movement within the island and improving links with the outside world. These include the development of Beane Field into Hewanorra Airport; the improvement of the East Coast Road; the extension of Castries Harbour; the new La Toc Road, and the Castries-Cap Highway.

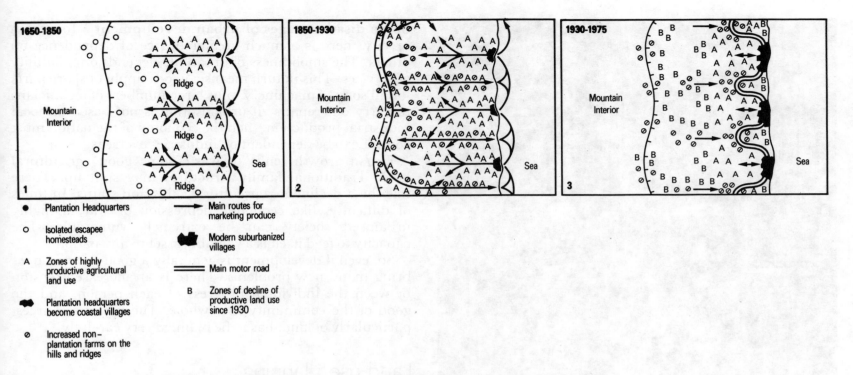

1650-1850 | Mountain Interior | Ridge | Ridge | Ridge | Sea

1850-1930 | Mountain Interior | Sea

1930-1975 | Mountain Interior | Sea

● Plantation Headquarters

○ Isolated escapee homesteads

A Zones of highly productive agricultural land use

◆ Plantation headquarters become coastal villages

⊘ Increased non-plantation farms on the hills and ridges

→ Main routes for marketing produce

◆ Modern suburbanized villages

═ Main motor road

B Zones of decline of productive land use since 1930

Fig. 32 Simplified view of land use and development in St Lucia since 1650

Urbanisation

One of the things that often accompanies economic development is a migration from the countryside to the towns. This movement is called urbanisation. Together with it we find two other things happening. One is urban growth; an increase in the area of built-up land. The other is urbanism; the adopting of an urban way of life. All three developments are very strong in Saint Lucia today. So it may be useful to consider briefly some of the advantages and disadvantages they bring to life on the island.

Some advantages of urban development Because, in towns, people are living close together it is easier to organise supplies of basic services. There is a better chance of getting piped water, electricity, and refuse collection services for every house. Then there is easier access to a greater variety of consumer goods, and to more opportunities for entertainment. There are bookshops and libraries, and generally a better educational environment exists in towns. There is easier access to medical services. Finally, there is a much greater range of occupations, giving, therefore, wider employment possibilities.

Some disadvantages of urban development   In a town or city there is a much greater degree of dependence on others. The smoothness of urban life depends on reliability of services. This in turn means a good supply of spare parts for all sorts of machinery, plus large numbers of technicians to carry out repairs. If urbanisation is not based on local industrial production, then the economy of the nation must continue to be dependent on industries overseas.

Urban growth tends to waste some good agricultural land. Traditional family structures may split up. There may be a decline in agricultural skills and status. In times of difficulty, like economic depression or war, a highly urbanised society can be extremely vulnerable. The capacity to feed itself is a valuable asset to any nation.

So, even if development is generally a good thing, it does bring many new problems. There is an increased tension between the individual progress of each person, and the good of the community as a whole. The use of resources, particularly of land, has to be planned very carefully.

Land-use planning

A regional or national scheme for land use in the future is called a structure plan, the object of which is to provide a framework for development. A structure plan for Saint Lucia might look like the map opposite.

In fact we can already see some of these developments on the ground. One example is the rapid urbanisation of the zone between Castries and Cap. This is an example of planning seeking to contain and organise ongoing developments. Another function of a structure plan is to promote a balanced development. This is why a number of alternative growth points have been suggested outside Castries. In particular there is the possibility of developing

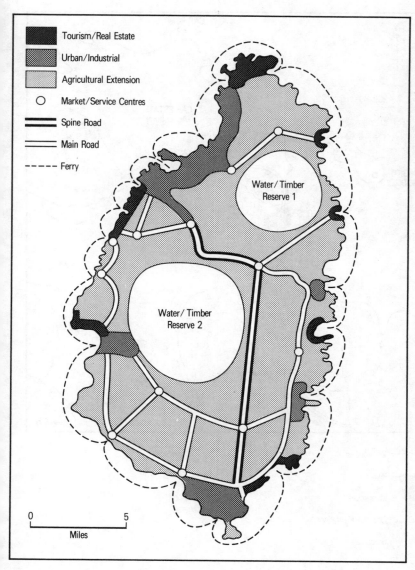

Fig. 33 A structure plan for St Lucia

Legend:
- Tourism/Real Estate
- Urban/Industrial
- Agricultural Extension
- O Market/Service Centres
- Spine Road
- Main Road
- --- Ferry

Water/Timber Reserve 1

Water/Timber Reserve 2

0 — 5 Miles

an industrial area near Vieux Fort, and supplementing land transport with ferry services. Thus the maritime locations of the major villages would regain their significance.

Structure plans are ideas. In order for development to take place on the ground, local surveys have to be made of the existing land-use. Investigation has to take place into the ownership of the land. If the Government wishes to develop land that is privately owned, it may purchase the land in the normal way, or it may be necessary to operate a compulsory purchase order. On the other hand, a private landowner may wish to develop his land. He will have to submit a planning proposal to the Government. The idea of this is to safeguard the interests of other members of the community. In other words to maintain the environment in the long term interests of the nation.

Scale & context in Saint Lucian development

Development is not only a question of local land-tenure and land-use, nor of a national plan. Saint Lucia is a small nation dependent for the foreseeable future on trade with, and aid from, the outside world.

Changes in Saint Lucia's relationships with other countries will obviously affect development prospects. For example, this nation was a member of the Caribbean Free Trade Area (CARIFTA), which has now developed into the Caribbean Community (CARICOM). In this way it is hoped that the small nations of the Caribbean may benefit from economies of scale. By forming a larger home market they may be able to lower the cost of living.

It has even been suggested by some regional planners that each member nation should specialise in one or two products for export within the community. For example Saint Lucia could become the copra producer. If this happened the whole context of development would be altered.

It is probably more likely that the Saint Lucian nation will continue to move towards economic diversification. The smallness of scale could be an advantage in that, together with the widespread individual land ownership traditional in the island, a genuinely mixed and balanced urban and rural life could develop. With better internal transport, instead of a division between 'Castries' and the 'Out-Districts', the whole island could become a sort of 'green suburb'.

Rodney Bay: a case study in development

As you can see from the map, there have been enormous changes in the Rodney Bay area which covers a large part of the north-west of Saint Lucia. In fact it is the location of one of the largest development projects in the Caribbean area. In addition to the official Rodney Bay Development Area there is the Cap Estate Development, and the village community of Gros Islet.

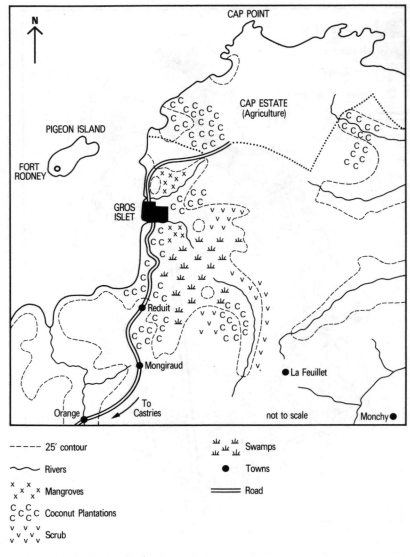

Fig. 34 Gros Islet Area in 1958

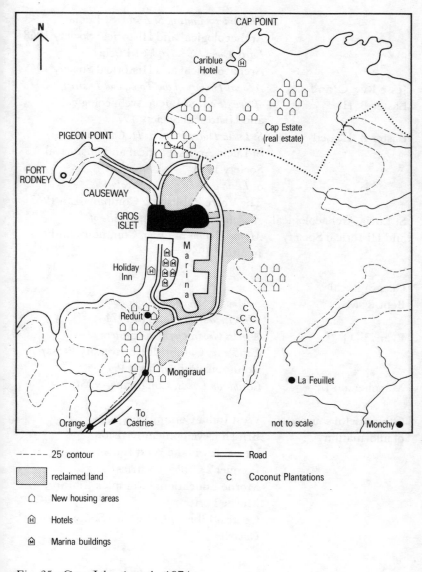

Legend:
----- 25' contour

▨ reclaimed land

⌂ New housing areas

ⓗ Hotels

ⓜ Marina buildings

═══ Road

C Coconut Plantations

Fig. 35 Gros Islet Area in 1974

Here a swamp is being transformed into a marina and tourist complex, an arid cattle ranch into a golf course and real estate development, and a declining fishing village into a modern suburban community.

Such an enormous project requires much careful planning and co-operation. It involves the Saint Lucia Government, the Caribbean Development Bank and companies from overseas. It shows how development in Saint Lucia is linked with the outside world.

Here is a young nation looking towards the future, but knowing that her agriculture and her people are still the most important resources for **survival** in the modern world.

Education and development

Much of the information for the Rodney Bay Case-Study comes from a survey made with Saint Lucian Junior Secondary School children in 1972 which was one of the case studies in *Social Studies in Practice* (see reading list). For development to succeed, the process must be fully understood by all the people concerned. So economic development in Saint Lucia goes hand-in-hand with educational development, welfare development and political development.

The nation of Saint Lucia is not the island, it is the people with whom this book was begun. Future development depends on the ideas and skills of the people. It depends on their understanding of the problems ahead, and ability to participate in solving them.

Learning about Saint Lucia This sort of book represents a small part of the development process. Social Studies is a part of curriculum development. It aims to provide a general background and introduction from which more specialist studies can proceed. It is hoped that those who read it may wish to learn more about the History, Geography, Economics, Politics and Sociology of the Island. There is a suggested reading list at the end of this book.

Key words

development	conservation	CARIFTA
urbanisation	CARICOM	land use
planning	networks	social stress
dependence	structure plan	import substitution
suburb	resources	local plans
participation	education	

Further reading

Ministry of Health and Education	*Development of St Lucia*, Lithographic Press 1971
Durham H, and Lewissohn F.	*St Lucia Tours and Tales*, Robertson Printing Corporation, New York 1971
Evans R.M.	*Social Studies in Practice*, Macmillan 1975
St Hill L.	*Manifesto One: The Land*, Letchworth Printing Press, Barbados

Historical

Jesse Rev. C,	*Outline of St Lucia's History*, St Lucia Archaeological and Historical Society 1956
	St Lucia: the romance of its place names, Archaeological and Historical Society 1966
	The Amerindians in St Lucia, St Lucia Archaeological and Historical Society 1968
	Early Days 1493-1763, St Lucia Archaeological and Historical Society 1969
Jesse Rev. C. and Easter B. H.	*A Short History of the Town and District of Veux Fort*, St Lucia Archaeological and Historical Society 1971
Easter B. H., (ed)	*St Lucia Diary of Lt. J. H. Caddy R.A.*, St Lucia Archaeological and Historical Society 1973
	St Lucia and the French Revolution, The Voice Publishing Company Ltd 1965
St Lucia Archaeological and Historical Society	*A Short Guide to the Historic site of Morne Fortune St Lucia*, Commonwealth Institute 1969

Geographical

Bent R. M.	*A Modern Secondary Geography of the West Indies*, Macmillan 1971
Evans F. C.	*A First Geography of the Eastern Caribbean*, Cambridge University Press/ Columbus Publications 1972
Macpherson J.	*Caribbean Lands*, Longmans 1973

Other useful sources of information	West India Committee
	British Development Division
	University of the West Indies
	Premier's Office, Castries
	Morne Educational Complex Library, Morne Fortune
	Central Library Columbus Square, Castries